Indoor Environmental Quality

Edited by Muhammad Abdul Mujeebu

Published in London, United Kingdom

IntechOpen

Supporting open minds since 2005

Indoor Environmental Quality
http://dx.doi.org/10.5772/intechopen.75787
Edited by Muhammad Abdul Mujeebu

Contributors
Naglaa Sami AbdelAziz, Pontip Nimlyat, John Anumah, Michael Odoala, Gideon Benjamin, Hany Hossameldin, Abdulrahman Abdullah Alshawan, Joseph Laquatra, Rui Pitarma, Gonçalo Miguel Santos Marques, Muhammad Abdul Mujeebu

First published in London, United Kingdom, 2019 by IntechOpen
IntechOpen is the global imprint of INTECHOPEN LIMITED, registered in England and Wales, registration number: 11086078, The Shard, 25th floor, 32 London Bridge Street
London, SE19SG – United Kingdom
Printed in Croatia

British Library Cataloguing-in-Publication Data
A catalogue record for this book is available from the British Library

Additional hard copies can be obtained from orders@intechopen.com

Indoor Environmental Quality
Edited by Muhammad Abdul Mujeebu
p. cm.
Print ISBN 978-1-78985-251-6
Online ISBN 978-1-78985-252-3

For EU product safety concerns:
IN TECH d.o.o., Prolaz Marije Krucifikse Kozulić 3, 51000 Rijeka, Croatia, info@intechopen.com or visit our website at intechopen.com.

Meet the editor

Dr. Muhammad Abdul Mujeebu obtained his Bachelor of Technology (BTech) degree in Mechanical Engineering from Kerala University, India; his Master of Technology (MTech) degree in Energy Management from the National Institute of Technology Calicut, India; and his PhD degree and Postdoctoral Fellowship from Universiti Sains Malaysia, Malaysia. He is currently serving as an associate professor at the College of Architecture and Planning, Imam Abdulrahman Bin Faisal University, Dammam, Saudi Arabia. His research areas include combustion in porous media, industrial cogeneration and trigeneration, microcombustors and MEMS, nanotechnology, electronic cooling and packaging, computational fluid dynamics analysis of energy systems, internal combustion engines and combustion, alternate fuels and refrigerants, solar cooling, indoor environmental quality, energy and thermal performance of buildings, and renewable energy systems. He has more than 75 publications in highly indexed journals.

Contents

Preface

First of all, I highly appreciate the initiative of IntechOpen to publish a book on indoor environmental quality (IEQ). This domain has attracted a large number of audiences and researchers, as evidenced by the dramatically increasing number of research works in the literature. Even though many books are available in this field, the open access facility of this book is expected to attract a wider readership.

Chapter 1 of this book provides an overview of IEQ, wherein the various constituents of IEQ are briefly explained, followed by research trends and concluding remarks. Chapter 2 presents an exclusive review of indoor air quality (IAQ), which is a major component of IEQ. Chapter 3 focuses on the monitoring of IAQ in healthy buildings. Chapter 4 deals with the guidelines to ensure the best illumination in indoor spaces. While Chapter 5 presents a practical study of IEQ in healthcare buildings. Chapter 6 deals with acoustic comfort in worship places. The authors of each chapter of this book are renowned experts in the field of IEQ. This book is expected to benefit undergraduate and postgraduate students, researchers, teachers, practitioners, policy makers, and every individual who has a concern for healthy life.

While presenting this book to an august audience, I gratefully acknowledge the contributions of the authors, the initiative of Ms. Anja Filipovic to launch the book, and the excellent administrative support of Ms. Dolores Kuzelj, all of whom collectively contributed to the successful release of this book.

Dr. Muhammad Abdul Mujeebu
Associate Professor,
Department of Building Engineering,
College of Architecture and Planning,
Imam Abdulrahman Bin Faisal University,
Dammam, Saudi Arabia

Chapter 1

Introductory Chapter: Indoor Environmental Quality

Muhammad Abdul Mujeebu

1. Overview

The term "indoor environmental quality" (IEQ) represents a domain that encompasses diverse sub-domains that affect the human life inside a building. These include indoor air quality (IAQ), lighting, thermal comfort, acoustics, drinking water, ergonomics, electromagnetic radiation, and many related factors [1], as depicted in **Figure 1**. Enhanced environmental quality can improve the quality of life of the occupants, increase the resale value of the building, and minimize the penalties on building owners.

IEQ in offices and other workplaces has a crucial role on the return on investment of businesses. A workplace with high IEQ obviously improves the workers' health and mood, thereby increasing their productivity. Therefore, the additional cost of maintaining high IEQ levels in workplaces will be paid back in a reasonable period and generates additional monetary returns thereafter. It should be noted that buildings being rated as "sustainable and green" do not truly guaranty their compliance with the desired IEQ level [2–5]. Therefore, IEQ should be given specific focus while designing new buildings as well as in building retrofit plans.

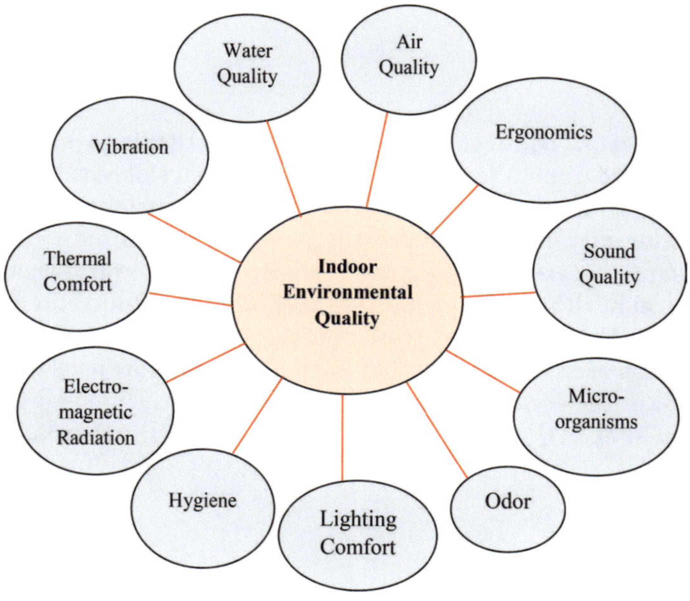

Figure 1.
IEQ components.

2. Indoor air quality

Indoor air quality (IAQ), which depends on airborne contaminants inside a building (or in a broader sense, any other enclosure such as a vehicle or an animal house), is one of the crucial factors that determine the quality of the indoor environment. Providing adequate air quality for the occupants is one of the most important functionalities of a building. Lung cancer (due to radon), Legionnaires' disease, carbon monoxide poisoning, allergy, and asthma are among the serious health implications of poor IAQ [6]. The "sick building syndrome" resulting from inadequate levels of IAQ significantly affects the health and productivity of office employees [7]. Though tremendous efforts are in progress to realize energy-efficient, green, and sustainable buildings, maintaining a safe level of IAQ in these buildings is an ongoing challenge. This is due to the fact that many energy-efficient measures in a building (such as reduced outdoor air ventilation rate, increased thermal insulation, and efficient cooling equipment) can have a detrimental impact on IAQ. Thus, alongside energy efficiency and sustainability, there has been a growing concern over air pollution inside buildings. Therefore, attempts to ensure energy efficiency and sustainability in buildings should simultaneously ensure enhanced health, comfort, and productivity of the occupants [6].

There are two major approaches to tackle IAQ issues in buildings: one is to increase the ventilation rate of outdoor air into the building, and the other is to minimize or control the sources of air pollution within and outside the building. Having said that, the first strategy would work only when the outdoor air is clean enough to improve IAQ [7]. The various sources that affect IAQ are, but not limited to, volatile organic compounds, biological pollutants, oxides of carbon and nitrogen, particulate matter, tobacco smoke, radon, mold, formaldehyde, pesticides, and combustion products. Heseltine and Rosen [8] outlined health issues associated with building moisture and biological agents, and the most important health problems identified are respiratory symptoms, allergies, asthma, and perturbation of the immunological system. A recent review [9] has revealed that carpets play a crucial role in IAQ, as they act as a sink for indoor air pollutants such as particles, allergens, and other biological pollutants.

3. Thermal comfort

The term "thermal comfort" refers to a condition that is governed by many environmental and human factors; in other words, physiological, physical, and sociopsychological factors. The environmental factors include air temperature, air velocity, humidity, radiant temperature, and relative humidity, while the major human factors are clothing and metabolic heat. The various other factors include physical health, mental condition, availability of food and drink, and acclimatization. This condition is mostly subjective, which cannot be directly quantified. It has been established that the thermal comfort level is acceptable if at least 80% of the occupants feel comfortable with it. Djongyang et al. [10] and Taleghani et al. [11] provided detailed insights into the thermal comfort in buildings.

4. Lighting comfort

Visible light falls in a narrow range in the electromagnetic spectrum, between ultraviolet and infrared wavelength ranges. Light has both particle and wave properties; when treated as a wave, light has a frequency that depends on the color of the struck surface. For instance, white surface reflects back most of the incident light, while a black surface absorbs most of it. The main aspects of lighting comfort are light level (intensity or brightness), contrast, and glare. The light intensity requirement depends on the type of activity in the building; for instance, operating rooms need a brighter level than living rooms. The term "contrast" refers to the ease of understanding or legibility; higher contrast gives higher clarity (e.g., black text on white paper provides the highest contrast). Glare is always undesirable as it causes a high level of discomfort in viewing the objects and affects the retina.

The visual comfort level is evaluated by means of some established glare metrics or indices; for example, glare probability (DGP) and daylight glare index (DGI) are used for assessing discomfort due to daylighting, while unified glare index (UGI), visual comfort probability (VCP), and CIE glare index (CGI) are employed for measuring the discomfort level of artificial lighting [12–15]. Several other indices are also available, as summarized by Carlucci et al. [12]. Galatioto and Beccali [16] reviewed the various aspects and concerns associated with the assessment of indoor daylighting.

5. Acoustic comfort

Building acoustics deals with controlling the quality of sound inside a building. It has two parts, namely, room acoustics and building acoustics, which deal with the sound propagation within a room and between rooms (through walls, doors, and floors), respectively. While the room acoustics focuses mainly on the sound quality (e.g., easy communication and high level of intelligibility in office spaces), the building acoustics is concerned with the "unsolicited" sound (e.g., the noise in a room should not be a nuisance to other rooms). The acoustic comfort in a building has a crucial impact on the health, well-being, communication, and productivity of the occupants. The acoustic comfort can be affected by factors such as the geometry and volume of a space, generation of sound within or outside the space, airborne noise transmission, impact noise, and the acoustic characteristics (absorption, transmission, and reflection of sound) of the interior surfaces. The measuring unit of sound intensity is decibels (dB), and of sound pitch is hertz (Hz). The comfortable range of sound for humans is typically 20–20,000 Hz.

The common parameters used for evaluating the acoustic performance of a building are reverberation time (RT), sound pressure level (SPL), early decay time (EDT), clarity (C_{50} for speech and C_{80} for music), sound definition or speech intelligibility (D or D50), and speech transmission index (STI). RT is defined as the time for the sound level to decay by 60 dB after a sound source has been switched off. EDT is similar to RT, but it is the initial rate of sound decay in a room, measured as the slope of a line 0–10 dB decay below the maximum sound level. D50 is defined as the ratio of the early received sound energy (0–50 ms after direct sound arrival) to the total received energy. Clarity is defined as the ratio of the energy in the early sound (received in the first 80 ms) to that in the reverberant sound. STI is a measure of speech transmission quality, which indicates the degree to which a transmission channel degrades speech intelligibility. STI ranges from 0 to 1; a speech transferred through a channel with STI of 1 is perfectly intelligible, but the intelligibility reduces as the STI approaches zero. International standards and guidelines (e.g., ISO 18233) are available for the measurement of these characteristics.

Extensive researches are in progress, on the acoustic comfort in buildings. In recent works, Tong et al. [17] studied the acoustical performance of classrooms and laboratories in a public school exposed to traffic environment, while Jeong et al. [18] focused on the acoustic design and evaluation of a concert hall. Tan et al. [19] introduced application of building information modeling to improve indoor acoustic performance. Few other studies include those reported by Lam et al. [20], Imran et al. [21], and Renterghem [22].

6. Ergonomics

Ergonomics deals with the design of objects, systems, and environment, in a manner that ensures human comfort. In fact, ergonomics encompasses all components of IEQ, simply because the prime objective of IEQ is human health and comfort. It covers diverse disciplines such as anatomy, physiology, psychology, and design. An indoor ergonomist should be specialized in the interrelationship between the human mind and body and the various aspects of a building such as architecture, interior design, building services, structure, materials, and microclimate. In general, environmental ergonomics deals with the interaction between people and their physical environment with particular importance on thermal comfort, lighting, noise, and vibration. Similar to ergonomics in a residential environment, ergonomics in offices and workplace is also a scientific discipline and a topic of research. Edmonds [23] defines the following factors that affect the workplace ergonomics: tasks, tools, equipment, area and space, environment, and organizational pattern. The Southeast Asian Network of Ergonomics Societies (SEANES) has introduced ergonomic checkpoints for indoor and outdoor workplaces for the purpose of motivating workers to recognize hazards in the work environment and adopt precautionary measures accordingly [24]. Similarly, Ushada et al. [25] developed environmental ergonomic control system for small and medium sized, by using worker workload and workstation temperature difference.

7. Electromagnetic field and radiation

Electromagnetic field is created by moving electric charges, microwaves, radio waves, electrical currents, and transformers. The low-frequency electromagnetic radiation prevailing mostly in indoors (due to electrical appliances, computers, wireless devices, etc.) can have detrimental effect on human health, and there are international regulations to deal with this problem (e.g., International Radiation Protection Association (IRPA)) [26]. Most of the regulations agree that exposure to electromagnetic field beyond the safe range of 0–300 Hz is harmful for the human body [27]. The possibility of health hazards such as acute lymphoblastic leukemia in children due to electromagnetic field exposure was well established decades ago [28] and continues to be a significant topic of research [26, 29–31].

8. Water quality

Adequate, safe, and accessible supply of drinking water is vital for the sustenance of human life especially in indoor environments where access to natural sources of water such as wells, ponds, rivers, and lakes is limited. Drinking water quality has a direct impact on human health. Infants, young children, weak and elderly people, and those who live in unhygienic environment are largely prone to waterborne deceases [32]. There is no universally applicable legislative framework for the implementation of standards to maintain drinking water quality. An approach that works in one country or region may not be suitable for other countries. Therefore, each country should develop its own legislation according to its requirements and capacity for implementation. However, while developing standards, the most common aspects that need to be taken into account are microbial safety, chemical safety, radiological safety, disinfection, and acceptability [32].

9. IEQ research trends

A huge number of literatures are available on the research on various aspects of IEQ, and a comprehensive review of these literatures is beyond the scope of this chapter. Many researchers have compiled them in their review articles [7, 33–42]. However, a brief overview of the exemplary researches is presented here. Most of the researches were on post-occupancy evaluation (POE) on IEQ of different types of common buildings (e.g., healthcare, office, educational, residential, etc.), through field measurements and user satisfaction surveys, while many other researchers were interested on POE of sustainable and green buildings. In these researches, the findings are usually compared with the prevailing local or global (as applicable) standards, and recommendations are made to address the issues identified.

9.1 IEQ of common buildings

Reynolds et al. [43] measured the physical, mechanical, and environmental factors affecting IEQ of office buildings in the United States (US). The measurements included endotoxin, total bioaerosols, and psychosocial parameters. Addressing the impact of IEQ on the occupant's productivity in offices, Kang et al. [44] investigated open-plan research offices in 19 Chinese universities by conducting survey on 231 subjects. The study identified five factors that significantly affected the office productivity, which are layout, air quality, thermal comfort, lighting, and acoustic comfort, where the acoustic comfort had the maximum impact. In a similar study [45], experiments were performed on the effect of indoor temperature on the IEQ user perception and productivity in office buildings, by choosing 9 females and 12 males. The parameters measured were air temperature, globe temperature, relative humidity, carbon dioxide (CO_2) concentration, and lighting and noise comforts. The indoor air temperature was varied by keeping the other IEQ parameters fixed. It was shown that the thermal environment had a significant impact on the thermal comfort and other IEQ factors. Kim et al. [46] focused on the impact of IEQ and work stress on the physiological responses of office workers and concluded that the most noticeable result of the experiment in this study is that a high CO_2 concentration and work stress could detrimentally influence the physiological and physiological responses, leading to abnormal variations in blood pressure. Similar studies on the effect of IEQ on office workers' performance are those reported by Haapakangas et al. [47], Suk [14], Zuo and Malone Beach [48], Ali et al. [49], Huang et al. [50], Frontczak et al. [51], Wong et al. [52], and Kosonen and Tan [53, 54].

Almeida and De Freitas [55] performed onsite measurements of temperature, relative humidity, CO_2 concentration, and ventilation rates in the classrooms of nine retrofitted and non-retrofitted school buildings in Portugal. The measurements were done during winter, mid-season, and summer conditions. In their observations, the non-retrofitted schools lack in the desired IEQ level, while retrofitted buildings did not have mechanical ventilation systems. Shan et al. [56] investigated the influence of indoor thermal condition and IAQ on students' health and performance through life cycle costing (LCC) approach, by considering two university classrooms. In the proposed LCC approach, metrics were defined for students' health (or well-being) and performance, which were subsequently translated into monetary values to quantify the impact of IEQ. The indicators considered for health and performance were sick leave and students' grade achievement, respectively. The findings of this study indicated the significance of incorporating students' health and performance into the design and operation of educational buildings. Few other researches focusing on educational buildings are those of Kim et al. [57], Vilčeková et al. [58], Jamaludin et al. [59], De Giuli et al. [60], and Nasir et al. [61].

Lai et al. [62] developed an IEQ assessment model for residential buildings in Hong Kong. The empirical model developed by using the data collected from 125 occupants from 32 residential buildings was useful to assess the acceptance level in terms of operative temperature, CO_2 concentration, and acoustic and lighting comforts. The study revealed that both thermal and acoustic comforts were the decisive contributors, while IAQ was the least. Huang et al. [63] studied the effect of IEQ of long-term care (LTC) facilities on the occupants' behavior, through survey. Garcia et al. [64] performed retrospective descriptive secondary analyses on the data collected (air exchange rates, temperature, and humidity) from indoor, outdoor, and personal air in residential buildings. Addressing the IEQ of healthcare buildings, Andrade et al. [65] performed user perception survey on hospital buildings in Portugal, considering physical and social aspects. De Giuli et al. [66] conducted survey and field measurements of three medical wards in a general hospital in Italy.

9.2 IEQ of sustainable and green buildings

As already mentioned, the IEQ level of sustainable and green buildings has been a concern of many researchers. Choi [67] proposed an explanatory model to understand the relationships among the occupants' perceptions on the IEQ level, overall facility, productivity, and sustainability ethic, in sustainable buildings. Hwang and Kim [68] performed post-occupancy evaluation (POE) of open offices in a Korean building that was certified as "1st Grade Building" Green. The studied parameters were indoor temperature, relative humidity, vertical temperature distribution, air velocity, predicted mean vote (PMV), radiant temperature, outdoor temperature, and humidity. Measurements were also done on the major indoor air contaminants, illuminance, and SPL. An online survey was also conducted among the occupants to know their perception on the IEQ level. The performance of this building was found to be satisfactory in terms of PMV and lighting, while it was weak for IAQ and acoustic comfort. Ravindu et al. [69] explored the IEQ level of a LEED-certified factory building in Sri Lanka, through questionnaire survey. They found that the building was performing low with regard to thermal comfort, ventilation, and ability to control indoor the environment. Altomonte et al. [3] studied the occupant satisfaction on IEQ in LEED- and BREEAM-certified office buildings and highlighted the importance of incorporating IEQ in the criteria for sustainable and green building certifications.

10. Concluding remarks

Indoor environmental quality is a very important scientific domain that deals with various aspects that govern the health, comfort, and productivity of the occupants and determine the value of a building. However, even though there is increasing awareness on the demand for sustainable, green, and high-performance buildings, ensuring the desired level of IEQ is often not given the deserving care. Consequently, most of the sustainable and green buildings lack in complying with the IEQ requirements. The building owners should rewrite their mindset to take into account the enormous potential for monetary returns and health benefits through improving the IEQ of the building. The following good practices are generally recommended to ensure a comfortable level of IEQ:

- Follow scientific practices of design, construction, renovation, operation, and maintenance, in compliance with the international standards.
- Adopt "source control" by minimizing the causes that lead to poor IEQ.
- Enhance the esthetics and indoor environment by proper integration of natural and man-made facilities.
- Minimize the dependence of artificial lighting and electrical equipment such as air conditioner, elevator, and fans, with a view to improve human health and minimize energy consumption.
- Ensure thermal comfort through proper design of the interior and microclimate.
- Facilitate proper ventilation and maintain acceptable air quality, by following standard guidelines.
- Adopt proper design and maintenance of HVAC system, and proper design and construction of the envelope, to prevent mold, fungi, airborne bacteria, and radon.
- Minimize the spread of pathogens by minimizing exposure to washrooms and by proper maintenance procedures.
- Avoid using products and materials, which contain harmful ingredients (such as formaldehyde) and produce harmful emissions.
- Ensure noise comfort and privacy, by suitably adopting the materials for walls, floors, and ceiling, and other standard means for acoustic comfort.
- Avoid unpleasant odors through selective use of products, regular and safe waste disposal, careful selection of cleaning products, isolation of contaminants, prohibition of smoking, and related measures.
- Establish a comfortable and healthy indoor lighting, through optimum integration of artificial and natural lightings, and use of energy-efficient, user-friendly, and eco-friendly artificial lighting.

- Maintain availability and accessibility of safe and clean drinking water in compliance with the water quality standards.
- Restrict and be aware of exposure to electromagnetic field and radiation, in the indoor environment.
- Ensure indoor ergonomic quality by providing ergonomic furniture and other facilities.
- Regularly conduct occupant surveys and post-occupancy evaluations.

Author details

Muhammad Abdul Mujeebu
Department of Building Engineering, College of Architecture and Planning, Imam Abdulrahman Bin Faisal University, Dammam, Saudi Arabia

*Address all correspondence to: mmalmujeebu@uod.edu.sa

References

[1] Almeida RMSF, de Freitas VP, Delgado JMPQ. School buildings rehabilitation-indoor environmental quality and enclosure optimization. In: Almeida RMSF, de Freitas VP, Delgado JMPQ, editors. School Buildings Rehabilitation. Switzerland AG: Springer Nature; 2015. pp. 5-17

[2] Spengler JD, Chen QY. Indoor air quality factors in designing a healthy building. Annual Review of Energy and the Environment. Amsterdam, Netherlands: Elsevier B.V; 2000;**25**:567-601

[3] Altomonte S, Saadouni S, Kent MG, Schiavon S. Satisfaction with indoor environmental quality in BREEAM and non-BREEAM certified office buildings. Architectural Science Review. 2017;**60**(4):343-355

[4] Steinemann A, Wargocki P, Rismanchi B. Ten questions concerning green buildings and indoor air quality. Building and Environment. 2017;**112**(2017):351-358

[5] Holmgren M, Kabanshi A, Sörqvist P. Occupant perception of 'green' buildings: Distinguishing physical and psychological factors. Building and Environment. 2017;**114**:140-147

[6] Persily AK, Emmerich SJ. Indoor air quality in sustainable, energy efficient buildings. HVAC & R Research. 2012;**18**(1-2):4-20

[7] Al horr Y, Arif M, Katafygiotou M, Mazroei A, Kaushik A, Elsarrag E. Impact of indoor environmental quality on occupant well-being and comfort: A review of the literature. International Journal of Sustainable Built Environment. 2016;**5**(1):1-11

[8] Heseltine E, Rosen J. WHO Guidelines for Indoor Air Quality: Dampness and Mould. Germany: World Health Organization Regional Office for Europe; 2009

[9] Becher R, Øvrevik J, Schwarze PE, Nilsen S, Hongslo JK, Bakke JV. Do carpets impair indoor air quality and cause adverse health outcomes: A review. International Journal of Environmental Research and Public Health. 2018;**15**(2):E184

[10] Djongyang N, Tchinda R, Njomo D. Thermal comfort: A review paper. Renewable and Sustainable Energy Reviews. 2010;**14**(9):2626-2640

[11] Taleghani M, Tenpierik M, Kurvers S, Van Den Dobbelsteen A. A review into thermal comfort in buildings. Renewable and Sustainable Energy Reviews. 2013;**26**:201-215

[12] Carlucci S, Causone F, De Rosa F, Pagliano L. A review of indices for assessing visual comfort with a view to their use in optimization processes to support building integrated design. Renewable and Sustainable Energy Reviews. 2015;**47**(7491):1016-1033

[13] Xue P, Mak CM, Huang Y. Quantification of luminous comfort with dynamic daylight metrics in residential buildings. Energy and Buildings. 2016;**117**:99-108

[14] Suk JY. Luminance and vertical eye illuminance thresholds for occupants' visual comfort in daylit office environments. Building and Environment. 2019;**148**:107-115

[15] Suk JY, Schiler M, Kensek K. Absolute glare factor and relative glare factor based metric: Predicting and quantifying levels of daylight glare in office space. Energy and Buildings. 2016;**130**:8-19

[16] Galatioto A, Beccali M. Aspects and issues of daylighting assessment:

A review study. Renewable and Sustainable Energy Reviews. 2016;**66**:852-860

[17] Tong Y, Abu Bakar H, Sari KM, Ewon U, Labeni M, Fauzan N. Effect of urban noise to the acoustical performance of the secondary school's learning spaces—A case study in Batu Pahat. IOP Conference Series: Materials Science and Engineering. 2017;**271**(012029):1-8

[18] Jeong K , Hong T, Kim SH, Kim J, Lee S. "Acoustic design of a classical concert hall and evaluation of its acoustic performance—A case study," Preprint, vol. 2018050309, no. May, pp. 1-12, 2018

[19] Tan Y, Fang Y, Zhou T, Wang Q, Cheng JCP, Kong H. Improve indoor acoustics performance by using building information modeling. In: 34th International Symposium on Automation and Robotics in Construction (ISARC 2017); 28 June-1 July 2017; Taipei, Taiwan. 2017

[20] Lam A, Hodgson M, Prodi N, Visentin C. Effects of classroom acoustics on speech intelligibility and response time: A comparison between native and non-native listeners. Building Acoustics. 2018;**25**(1):35-42

[21] Imran M, Heimes A, Vorländer M. Auralization of airborne sound transmission for coupled rooms in virtual reality. In: 44th Annual Conference on Acoustics (DAGA 2018); 19-22 March 2018; Munich, Germany. 2018

[22] Van Renterghem T. Improving the noise reduction by green roofs due to solar panels and substrate shaping. Building Acoustics. 2018;**25**(3):219-232

[23] Edmonds J. Overview of human factors engineering. In: Human Factors in the Chemical and Process Industries. Amsterdam, Netherlands: Elsevier B.V., Elsevier Inc.; 2016. pp. 153-167

[24] Khalid H, Kogi K, Helander M. Ergonomics intervention of workplaces using SEANES ergonomic checkpoints. In: 20th Congress of the International Ergonomics Association (IEA 2018); 2018. pp. 1125-1134

[25] Ushada M, Suyantohadi A, Khuriyati N, Okayama T. Identification of environmental ergonomics control system for Indonesian SMEs. In: 3rd International Conference on Control, Automation and Robotics; 24-26 April 2017; Nagoya, Japan. 2017. pp. 453-456

[26] Fayed I, Bedda M. Electromagnetic radiation and its effects on human beings: Survey and environmental recommendations. In: 15th Scientific Symposium for Hajj, Umrah and Madinah visit—Scientific Portal for 1436AH; 27-28 May 2015; Saudi Arabia. 2015. pp. 35-47

[27] Valberg PA, van Deventer TE, Repacholi MH. Workgroup report: Base stations and wireless networks--radiofrequency (RF) exposures and health consequences. Environmental Health Perspectives. 2007;**115**(3):416

[28] Linet MS, Hatch EE, Kleinerman RA, Robison LL, Kaune WT, Friedman DR, et al. Residential exposure to magnetic fields and acute lymphoblastic leukemia in children. The New England Journal of Medicine. 1997;**337**(1):1-8

[29] Kheifets L, Repacholi M, Saunders R, Van Deventer E. The sensitivity of children to electromagnetic fields. Pediatrics. 2005;**116**:303-313

[30] Verloock L, Joseph W, Vermeeren G, Martens L. Procedure for assessment of general public exposure from WLAN in offices and in wireless sensor network testbed. Health Physics. 2010;**98**(4):628-638

[31] Bhatt CR, Redmayne M, Billah B, Abramson MJ, Benke G. Radiofrequency-electromagnetic

field exposures in kindergarten children. Journal of Exposure Science & Environmental Epidemiology. 2017;**27**(5):497-504

[32] WHO. Guidelines for Drinking-Water quality: First Addendum to Third Edition, Vol 1—Recommendations. Geneva, Switzerland: World Health Organization; 2006

[33] Bluyssen PM, Janssen S, van den Brink LH, de Kluizenaar Y. Assessment of wellbeing in an indoor office environment. Building and Environment. 2011;**46**(12):2632-2640

[34] Yu CWF, Jeong TK. Building environmental assessment schemes for rating of IAQ in sustainable buildings. Indoor and Built Environment. 2011;**20**(1):5-15

[35] Kim J, de Dear R, Cândido C, Zhang H, Arens E. Gender differences in office occupant perception of indoor environmental quality (IEQ). Building and Environment. 2013;**70**:245-256

[36] Sakhare VV, Ralegaonkar RV. Indoor environmental quality: Review of parameters and assessment models. Architectural Science Review. 2014;**57**(2):147-154

[37] Rashid M. A review of the empirical literature on the relationships between indoor environment and stress in health care and office settings—Problems and prospects of sharing evidence. Environment and Behavior. 2015;**40**(2):21

[38] Nimlyat PS, Kandar MZ. Appraisal of indoor environmental quality (IEQ) in healthcare facilities: A literature review. Sustainable Cities and Society. 2015;**17**:61-68

[39] Al Horr Y, Arif M, Kaushik A, Mazroei A, Katafygiotou M, Elsarrag E. Occupant productivity and office indoor environment quality: A review of the literature. Building and Environment. 2016;**105**:369-389

[40] Kolokotsa D, Santamouris M. Review of the indoor environmental quality and energy consumption studies for low income households in Europe. Science of the Total Environment. 2015;**536**:316-330

[41] Sarbu I, Sebarchievici C. Aspects of indoor environmental quality assessment in buildings. Energy and Buildings. 2013;**60**:410-419

[42] Heinzerling D, Schiavon S, Webster T, Arens E. Indoor environmental quality assessment models: A literature review and a proposed weighting and classification scheme. Building and Environment. 2013;**70**:210-222

[43] Reynolds SJ, Black DW, Borin SS, Breuer G, Burmeister LF, Fuortes LJ, et al. Indoor environmental quality in six commercial office buildings in the Midwest United States. Applied Occupational and Environmental Hygiene. 2010;**16**(11):1065-1077

[44] Kang S, Ou D, Mak CM. The impact of indoor environmental quality on work productivity in university open-plan research offices. Building and Environment. 2017;**124**:78-89

[45] Geng Y, Ji W, Lin B, Zhu Y. The impact of thermal environment on occupant IEQ perception and productivity. Building and Environment. 2017;**121**:158-167

[46] Kim J, Kong M, Hong T, Jeong K, Lee M. Physiological response of building occupants based on their activity and the indoor environmental quality condition changes. Building and Environment. 2018;**145**:96-103

[47] Haapakangas A, Hallman DM, Mathiassen SE, Jahncke H. Self-rated productivity and employee well-being in activity-based offices: The

role of environmental perceptions and workspace use. Building and Environment. 2018;**145**:115-124

[48] Zuo Q, Malone Beach EE. Assessing staff satisfaction with indoor environmental quality in assisted living facilities. Journal of Interior Design. 2017;**42**(1):67-84

[49] Ali AS, Chua SJL, Lim MEL. The effect of physical environment comfort on employees' performance in office buildings: A case study of three public universities in Malaysia. Structural Survey. 2015;**33**(4-5):294-308

[50] Huang L, Zhu Y, Ouyang Q, Cao B. A study on the effects of thermal, luminous, and acoustic environments on indoor environmental comfort in offices. Building and Environment. 2012;**49**(1):304-309

[51] Frontczak M, Schiavon S, Goins J, Arens E, Zhang H, Wargocki P. Quantitative relationships between occupant satisfaction and satisfaction aspects of indoor environmental quality and building design. Indoor Air. 2012;**22**(2):119-131

[52] Wong LT, Mui KW, Hui PS. A multivariate-logistic model for acceptance of indoor environmental quality (IEQ) in offices. Building and Environment. 2008;**43**(1):1-6

[53] Kosonen R, Tan F. Assessment of productivity loss in air-conditioned buildings using PMV index. Energy and Buildings. 2004;**36**(10):987-993

[54] Kosonen R, Tan F. The effect of perceived indoor air quality on productivity loss. Energy and Buildings. 2004;**36**(10):981-986

[55] Almeida RMSF, De Freitas VP. Indoor environmental quality of classrooms in Southern European climate. Energy and Buildings. 2014;**81**:127-140

[56] Shan X, Melina AN, Yang EH. Impact of indoor environmental quality on students' wellbeing and performance in educational building through life cycle costing perspective. Journal of Cleaner Production. 2018;**204**:298-309

[57] Kim J, Hong T, Jeong J, Lee M, Lee M, Jeong K, et al. Establishment of an optimal occupant behavior considering the energy consumption and indoor environmental quality by region. Applied Energy. 2017;**204**:1431-1443

[58] Vilčeková S, Kapalo P, Mečiarová Ľ, Burdová EK, Imreczeová V. Investigation of indoor environment quality in classroom-case study. Procedia Engineering. 2017;**190**:496-503

[59] Jamaludin NM, Mahyuddin N, Akashah FW. Assessment of indoor environmental quality (IEQ): Students well-being in University classroom with the application of landscaping. MATEC Web of Conferences. 2016;**66**:00061

[60] De Giuli V, Da Pos O, De Carli M. Indoor environmental quality and pupil perception in Italian primary schools. Building and Environment. 2012;**56**:335-345

[61] Nasir ARM, Musa AR, Che-Ani AI, Utaberta N, Abdullah NAG, Tawil NM. Identification of indoor environmental quality (IEQ) parameter in creating conducive learning environment for architecture studio. Procedia Engineering. 2011;**20**:354-362

[62] Lai ACK, Mui KW, Wong LT, Law LY. An evaluation model for indoor environmental quality (IEQ) acceptance in residential buildings. Energy and Buildings. 2009;**41**(9):930-936

[63] Huang YC, Chu CL, Chang Lee SN, Lan SJ, Hsieh CH, Hsieh YP. Building

users' perceptions of importance of indoor environmental quality in long-term care facilities. Building and Environment. 2013;**67**:224-230

[64] Garcia F, Shendell DG, Madrigano J. Relationship among environmental quality variables, housing variables, and residential needs: A secondary analysis of the relationship among indoor, outdoor, and personal air (RIOPA) concentrations database. International Journal of Biometeorology. 2017;**61**(3):513-525

[65] Andrade C, Lima ML, Fornara F, Bonaiuto M. Users' views of hospital environmental quality: Validation of the perceived hospital environment quality indicators (PHEQIs). Journal of Environmental Psychology. 2012;**32**(2):97-111

[66] De Giuli V, Zecchin R, Salmaso L, Corain L, De Carli M. Measured and perceived indoor environmental quality: Padua Hospital case study. Building and Environment. 2013;**59**:211-226

[67] Choi S. The relationships among indoor environmental quality, occupant satisfaction, work performance, and sustainability ethic in sustainable buildings [PhD thesis]. The University of Minnesota; 2011

[68] Hwang T, Kim JT. Assessment of indoor environmental quality in open-plan offices. Indoor and Built Environment. 2013;**22**(1):139-156

[69] Ravindu S, Rameezdeen R, Zuo J, Zhou Z, Chandratilake R. Indoor environment quality of green buildings: Case study of an LEED platinum certified factory in a warm humid tropical climate. Building and Environment. 2015;**84**:105-113

Chapter 2

Indoor Air Quality

Joseph Laquatra

Abstract

Indoor air pollution is an international health concern because people spend a majority of their time indoors. Children are at a higher risk of health problems from pollutant exposure, especially because air in the child breathing zone is more polluted than it is in the adult breathing zone. Pollutants of concern include biological contaminants, combustion pollutants, volatile organic compounds, and radon and other soil gases. Humans have a history with lead and asbestos that goes back thousands of years to the ancient Romans and Egyptians. These two pollutants are still problems in older homes and apartments. All of these toxicants can be minimized or abated. Awareness of these issues is a critical first step in improving air quality in places where people live.

Keywords: biological contaminants, combustion pollutants, volatile organic compounds, radon, lead, asbestos, child breathing zone

1. Introduction

In recent years, indoor air pollution has become an international health concern. Research has shown that people spend about 90% of their time indoors [1] and 75% of their time indoors in their homes [2]. Some people such as children, the elderly, and infirm spend most or all of their time indoors [3, 4]. Research also indicates that pollutant levels can be higher indoors than outdoors [5]. Concerns about indoor air quality have led to indoor air management becoming a new consumer skill. Steps involved in indoor air management include identifying a pollutant of concern, controlling it at its source, and if that fails, mitigation. Residential indoor air pollutants include biological contaminants, volatile organic compounds, radon and other soil gases, combustion pollutants, lead, and asbestos.

2. Biological contaminants

Biological contaminants include mold, viruses, bacteria, pollen, animal dander, and dust mites. Moisture plays an essential role in the presence of biological contaminants. As shown in **Figure 1**, warm air holds more water vapor than cold air. The cube on the left represents a volume of air that is at 75°F, with 30% relative humidity. This means that it is holding 30% of the moisture that it is *capable* of holding. When that same amount of air cools to 40°F, it contains the *same amount* of water, but it is now at 100% relative humidity. In other words, it is holding all of the moisture that it *can* hold. Moisture will condense at 100% relative humidity. This is also called the saturation point or the dew point temperature.

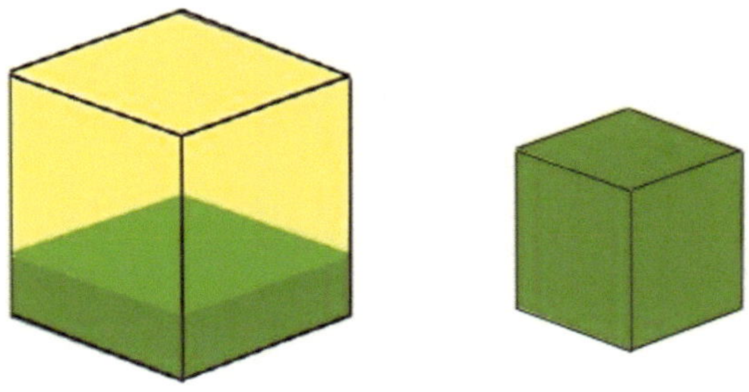

Figure 1.
Warm air holds more moisture than cold air.

When warm, moist air comes in contact with a cold surface, the water vapor in that air condenses to liquid water. In the case of a cold window, when warmer, humid air moves closer to the window, its temperature drops, and therefore its moisture-holding capability also drops. When this air touches the window, it condenses. The same thing happens on a warm and humid summer day, when warm, humid air condenses on cold beverage bottles, cans, or glasses. Sometimes, condensation on a window can be a nuisance. Other times, it can be serious enough that moisture will accumulate on the sash and on the sill, causing mold growth, warping that will damage the airtight seal between panes of glass, and even rotting. Mold spores are ubiquitous, and when a spore lands on a surface at the right temperature, with a food source—in this case, cellulose—and moisture, mold will grow.

Mold is a fungus; and as fungi grow, they release large numbers of spores into the air. And as mold digests cellulosic products, such as wood, as food, it releases carbon dioxide, water, and microbiological volatile organic compounds (mVOCs) into the air. Airborne spores affect asthmatics and people with allergies by acting as asthma triggers and the cause of respiratory illness. Microbiological volatile organic compounds are responsible for the musty smell associated with mold growth. Inhalation of mVOCs by humans can cause mycotoxicosis, symptoms of which include difficulty breathing, sore throat, bloody nose, and skin rashes.

Preventing health problems caused by exposure to mold is done by controlling moisture in homes. This means maintaining relative humidity at levels that do not allow for moisture condensation on windows and other surfaces, regularly inspecting plumbing pipes and fixtures for leaks, and preventing the entry of water from outside the home by maintaining roofs and siding and having a water-managed foundation.

Another biological contaminant commonly found in homes is the house dust mite, which feeds on skin cells that are naturally shed from human bodies. Fecal pellets from this microscopic arachnid contain a protein that is an allergen and asthma trigger. Dust mites thrive in humid environments and live in upholstered furniture, bedding, carpeting, and stuffed animals. They cannot survive at relative humidity levels below 50% [6]. Other biological contaminants in indoor air include viruses, bacteria, pollen, and animal dander. All of these can be controlled through regular house cleaning.

2.1 Ventilation

A number of factors contribute to the high levels of energy efficiency that are now possible in new and existing homes. Airtightening measures—those that prevent air infiltration through the building shell—are among the most critical of these. In new construction and in the improvement of an existing home, low air infiltration rates are achieved through an attention to the details of both construction materials and practices. And as air leakage has decreased in homes, ventilation has become a residential design issue because of problems that arise from excess moisture and other indoor air pollutants.

Before airtightening measures were as widespread as they now are, ventilation of homes was achieved naturally, as air leaked in and out of cracks in the building shell—around windows and doors, where dissimilar building materials meet, and other places. Natural ventilation is undesirable because it can never be controlled. Its rate depends on wind speed, vegetation around a house, site topography, and other variables. And natural ventilation imposes large energy costs on a home because the incoming infiltration air must be heated in the winter in cold climates. But in the absence of natural ventilation, mechanical ventilation is necessary for removing moisture and other pollutants as well as bringing fresh air into a home.

A basic mechanical ventilation system consists of exhaust fans, which are ducted to the outdoors, in kitchens and bathrooms. Conventional clothes dryers should always be ducted to the outdoors, although some electric clothes dryers vent into the washer. And some clothes washers also act as dryers. An issue that arises in airtight homes is the provision of make-up air for exhaust systems. As exhaust fans pull air out of a house, that air must be replaced. In a leaky house, that air is supplied through infiltration. This happens because the fans place negative pressure on a house and, if no windows are open, pull in air from cracks that exist in the building enclosure or from a chimney, which can be dangerous if the chimney is connected to an operating combustion appliance. Other ventilation systems exist that not only pull air out of a house but also provide make-up air.

Figure 2 shows temperature-difference-driven infiltration, also called the stack effect. In simpler terms: a house comes under negative pressure as warm air naturally rises to upper levels of a house. That warm air escapes through various faults in the building enclosure, including cracks that form at junctions of different types of construction materials, such as those where brick meets wood siding. Warm air

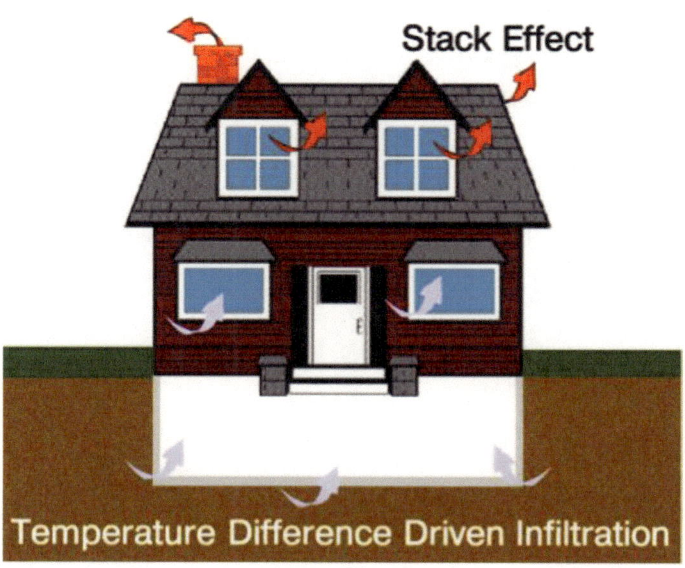

Figure 2.
The stack effect.

also escapes from unsealed cracks around windows and doors. All air that leaves a house in this manner must be replenished. This happens when air leaving the house creates suction pressure on lower house levels, which causes soil gases, including radon, to be pulled into the house.

Figure 3 shows how combustion appliances can also bring a house under negative pressure. All combustion appliances use some type of fuel, whether it is fuel oil, natural gas, propane, or wood. Oxygen is needed to fuel the fire, and if that oxygen comes from indoor air, it will put negative pressure on a house, just like the stack effect does. Air gets drawn into the appliance, fuels the fire, and that air needs to be replaced. The replaced air comes in through cracks in the building enclosure as well as cracks in the foundation of the basement, which can allow soil gases to enter the home.

A solution to negative pressure caused by a combustion appliance is to use a sealed combustion appliance. This type of furnace or boiler brings air to the combustion chamber through a pipe that originates outside the house. Sealed systems typically have a second heat exchanger that extracts heat from combustion gases that would normally be exhausted by the chimney in a conventional system. Instead, extracting additional heat from combustion gases results in exhaust gases that are cool enough to be exhausted from the house through a pipe through an exterior wall, much like a clothes dryer vent. Because the combustion air comes from outside the house, the building does not come under negative pressure.

Approaches to residential ventilation can be categorized as exhaust, supply, and balanced systems. Fans that pull air out of a space such as a bathroom exhaust fan or a kitchen range ventilation hood comprise basic exhaust ventilation systems that most people are familiar with. As noted above, however, these fans can place an airtight house under negative pressure.

Variations of exhaust systems provide make-up air to the house in some manner. The simplest way to do this is to install passive vents, which are small, screened openings in exterior walls. These admit air by opening when the home comes under negative pressure, such as when an exhaust fan is turned on. Passive vents are only recommended for use in very small, airtight homes in which depressurization is safe. Home depressurization is safe if all combustion appliances receive combustion air from outside the home; there are no fireplaces in the home; the home has no attached garage; and the home is not located in a high radon area.

More commonly used than exhaust fans with passive vents is a central exhaust system that pulls air out of a house combined with a fan that pulls fresh air into the house and delivers it through ducts to individual rooms, usually each bedroom

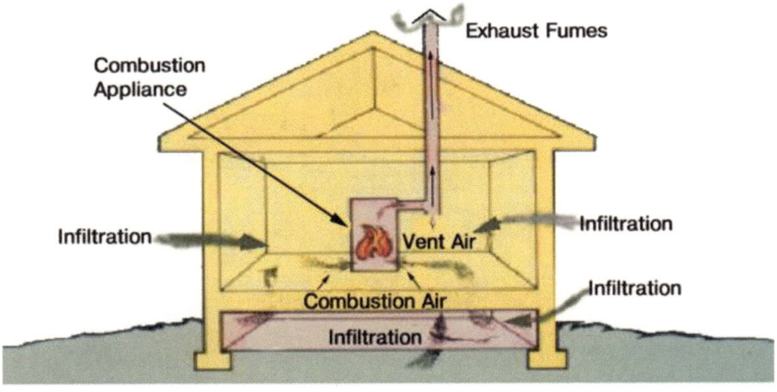

Figure 3.
Combustion air concepts.

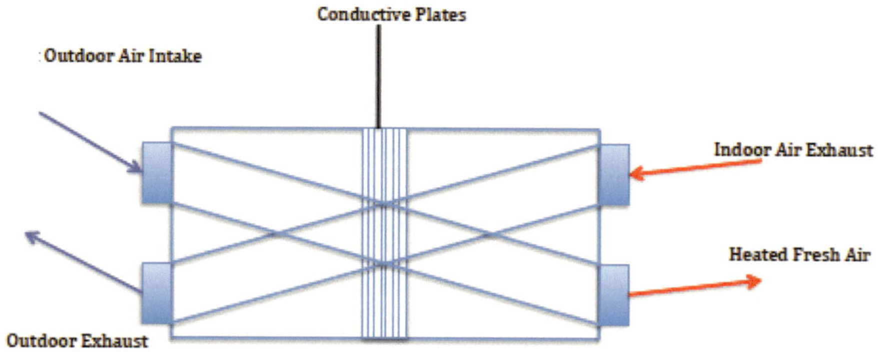

Figure 4.
Heat recovery ventilator.

and living area. Whole-house fans are effective in this type of supply system. A variation of this system, if the house has a forced air furnace, is to deliver outdoor air to the return duct, so that it can be mixed with indoor air and heated before it is delivered to the rooms.

A heat recovery ventilator (HRV)—also referred to as an air-to-air heat exchanger—is a balanced system that consists of a device which pulls fresh air into a home at the same time that it is exhausting air out of the home. As seen in **Figure 4**, the two airstreams are separated but pass over a core of conductive plates or heat exchanger that transfers heat from the warmer airstream to the colder one. A heat recovery ventilator also dehumidifies the home, because the warmer airstream contains moisture that condenses during the exchange process. The resulting water is delivered to a drain through a tube. HRVs can be stand-alone units with ducts or they can be integrated with the ducts of a forced air furnace. In addition to the basic systems described above, other variations exist, including central exhaust/supply systems with dehumidification and systems with air filtration options. Several studies have analyzed the cost effectiveness of various ventilation systems by examining purchase and installation costs, annual operating costs, and additional imposed heating costs (to heat incoming air). In addition to costs, benefits that are difficult to quantify include increased human comfort and the prevention of moisture problems.

Taylor et al. [7] examined the cost-effectiveness of heat recovery ventilators and concluded that these units provide positive life-cycle cost savings throughout much of the United States, although not in the colder, northern tier states.

The International Residential Code (IRC) specifies mechanical ventilation standards for new homes, which vary depending on the size of the house, number of bedrooms, and tested air infiltration rate [8]. The infiltration rate is measured with a blower door test, a specialized piece of equipment that measures a home's air change per hour (ACH). ACH measures the extent to which outdoor air leaks into homes through cracks around windows, doors, and where dissimilar building materials meet. An airtight home has a low ACH; a leaky, drafty home has a high ACH.

3. Volatile organic compounds

Volatile organic compounds (VOCs) are gases released from some solids or liquids at room temperature. Many VOCs found in household air have adverse health impacts, including eye, nose, and throat irritation; asthma exacerbation; lung, kidney, and central nervous system damage; and cancer [9]. VOC sources include building products, paints, strippers, solvents, wood preservatives, air fresheners, hobby supplies, pesticides, dry-cleaned clothing, and more.

The World Health Organization (WHO) categorizes VOCs by the ease with which they are emitted from materials and uses the terms very volatile organic compounds (VVOC), volatile organic compounds (VOCs), and semivolatile organic compounds (SVOCs) [10]. As mentioned earlier, VOCs produced by mold are referred to as microbiologic volatile organic compounds (mVOCs). But all of these fall into the broad category of VOCs.

Formaldehyde, a colorless, strong-smelling gas, is a common VOC used in the production of building materials, cabinets, furnishings, household cleaners, paints, landscape materials, and other products. It is used in the production of plywood, particle board, and medium density fiberboard. Formaldehyde is released into the air in a process referred to as off-gassing. Formaldehyde is also a component of cigarette smoke and a combustion product of wood, kerosene, natural gas, oil, and gasoline.

Adverse health effects from formaldehyde exposure include eye, nose, and throat irritation; wheezing and coughing; and allergic reactions. Long-term exposure to high levels of formaldehyde can cause cancer in humans.

Other VOCs of concern in indoor air include benzene, styrene, xylene, and methylene chloride. Benzene is a human carcinogen that is present in environmental tobacco smoke, solvents, plywood, particle board, fiberglass, wood paneling, adhesives, paint, caulking, and wood strippers. Styrene is used in the manufacturing of plastics, rubber, food containers, carpet backing, vinyl flooring, and resins. Acute health effects from styrene exposure include mucous membrane irritation; depression; muscle weakness; and eye, nose, and throat irritation. Chronic effects include hearing loss, peripheral neuropathy, and kidney damage. Xylene is a solvent and is a component of rubber and adhesives. Health effects from exposure include depression of the central nervous system, dizziness, irritability, and vomiting. Methylene chloride, which is also known as dichloromethane, is used in paint, paint strippers, and adhesives. Exposure can cause damage to the central nervous system, liver cancer, and lung cancer. This is not an exhaustive list of VOCs found in homes but is meant to illustrate potential hazards from common materials.

3.1 VOCs and safety

When using any product that contains VOCs, provide adequate ventilation to the work area, meet or exceed any label precautions, buy in quantities that will be consumed quickly, and dispose of containers safely. Do not allow children or pets to become exposed to these products. Low-VOC- and No-VOC-containing products are becoming widely available. When possible, use these products instead of conventional alternatives.

4. Radon

Radon is a radioactive gas that has no odor, taste, or color. It is produced during the decay of uranium, has a half-life of 3.8 days, and emits alpha and gamma radiation [11]. Uranium exists in soils all over the world. The radioactive decay process causes uranium to decay to uranium. Uranium and radium are solid elements. But radium decays to a gas: radon. Radon moves easily through permeable soils, such as gravelly and sandy soils, than it does through impermeable soils, such as clay [12]. Cracks in a house foundation and other openings, such as those around pipes that penetrate a house foundation, serve as radon pathways into the house. Radon continues in the decay process once it is inside a home. Radon's decay products are lead, polonium, and bismuth. These decay products become attached to microscopic particulates in house air, which are inhaled by people in the house and lead to lung cancer.

The process through which radon enters a home is displayed in **Figure 5**. Radon is the second-leading cause of lung cancer after cigarette smoking; radon exposure is responsible for 21,000 deaths per year in the USA [13]. Between one and seven percent of lung cancer fatalities in the USA have been attributed to radon exposure [13].

Radon's presence can be confirmed through the use of short- or long-term radon detectors. A short-term detector consists of activated carbon, which adsorbs (collects on the surface of carbon granules) radon, is inexpensive and simple to use. Once activated, it is placed in the lowest room of the house and kept in place for three days. The house should be tested under closed house conditions. This means all windows are closed for the duration of the test, and doors are used only for normal entrances and exits. After the test period, the detector is sent to a laboratory for analysis. The laboratory then reports the test results to the sender. Radon levels in a house vary over time because of changes in weather and atmospheric pressure. So, a short-term test is effectively a snapshot of radon levels at a particular time. A long-term radon test uses what is known as an alpha track detector. This is placed in a home's living room and stays there for 90 days to a year. This type of test provides a better result of a home's radon level.

The U. S. Environmental Protection Agency (EPA) recommends that mitigation systems be installed at or above the Action Level of 4 picocuries per liter (pCi/L) of air [14]. A mitigation system for an existing home consists of a PVC pipe that is installed through the floor of the lowest level of a home, often a basement, into a layer of gravel. That pipe is carried up through the house attic and through the roof. This pipe can also be installed on a house exterior wall. Often, an inline exhaust fan is connected to the PVC pipe and is used to pull soil gas from below the house. When that fan is used, the mitigation system is referred to as an active system. An inline fan is not always necessary and a passive mitigation system is used instead.

The EPA has developed a U.S. map that designates counties into zones. In EPA-designated Zone 1 counties, indoor radon levels are expected to be 4 pCi/L or higher; houses in Zone 2 counties are expected to have radon levels between 2 and 4 pCi/L; homes in Zone 3 counties are expected to have radon levels below 2 pCi/L.

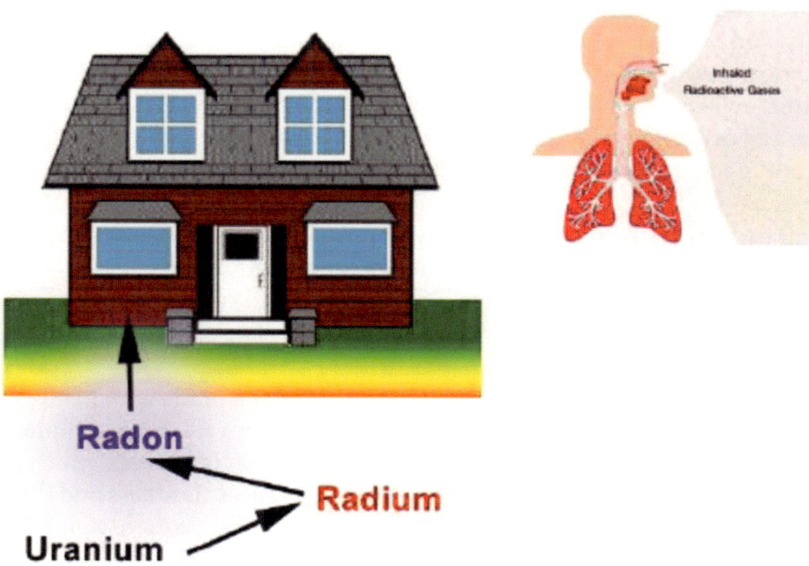

Figure 5.
Radon entering a home.

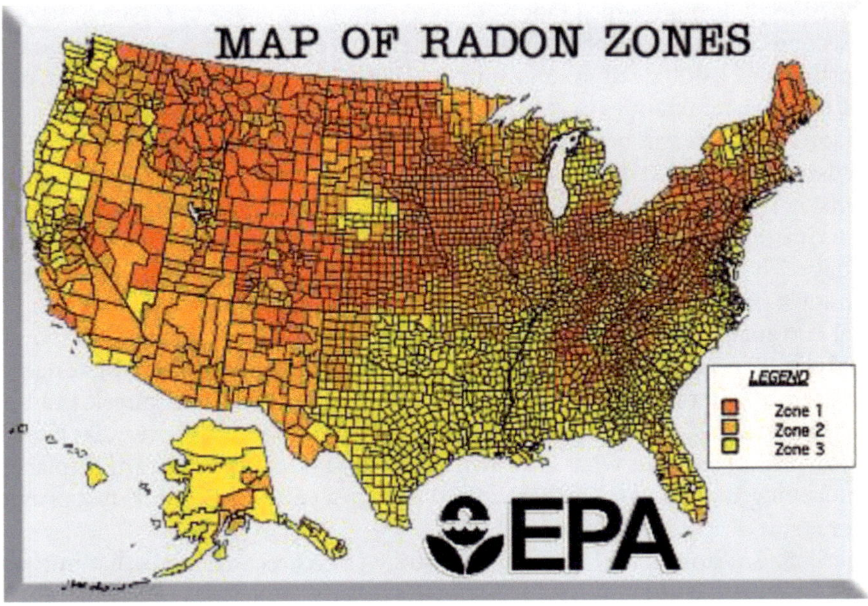

Figure 6.
EPA Radon Zone Map.

Figure 6 shows the EPA Radon Map. Zone 1 counties are red; Zone 2 counties are orange; and Zone 3 counties are yellow.

Radon-resistant construction techniques are recommended for new homes built in Zone 1 counties [15].

5. Combustion products

Combustion products comprise another category of indoor air pollutants. They consist of nitrogen oxides, sulfur dioxide, carbon monoxide, respirable particulates, and water. Nitrogen oxides, sulfur dioxides, and respirable particulates are lung irritants, and carbon monoxide (CO) can kill. To avoid indoor these pollutants, combustion-based, unvented space heaters should not be in the home. Central heating system systems should be regularly serviced: annual servicing for fuel oil-based systems and every 2 years for gas systems. Smoking should not be permitted in a home, and a gas kitchen range should have an exhaust fan over it that is vented to the outside.

Combustion products can pollute the air inside a home when components of a central heating system are damaged and leak combustion gases into indoor air. Indoor use of combustion-based electric generators will also do this. And when a house comes under negative pressure, combustion gases can be drawn from a chimney or fireplace into the home.

Normally, when a person breathes healthy air, oxygen binds with hemoglobin in blood to form oxyhemoglobin. When a person breathes air that is polluted with CO, CO binds with hemoglobin, and carboxyhemoglobin is formed, which prevents oxygen from getting to the brain. At low levels, this causes tiredness and dizziness. At higher levels, gradual suffocation and death occur. Carbon monoxide is responsible for hundreds of deaths and thousands of emergency room visits in the USA per year [16]. These are all preventable deaths and often occur when people are sleeping. Every home and apartment should have at least one carbon monoxide detector installed in the hallway outside the sleeping area. Carbon monoxide detectors are important, but they are no substitute for regular servicing of combustion appliances and common sense safety procedures with combustion appliances in the home.

6. Lead

People have used lead for numerous purposes for at least 7000 years [17]. Before 1550 BCE, ancient Egyptians used lead as a medicine and for decorative purposes. When babies were born, a lead ball was placed on their belly buttons to stop bleeding. Women would decorate their nipples with lead and breast feed their babies. People have been aware of lead poisoning for over 2000 years [17]. In spite of this, lead was not banned as an ingredient in residential paint in the USA until 1978 and in gasoline until 1986. There are an estimated 24 million homes and apartments in the USA with lead-based paint [18].

Negative health impacts from lead exposure include reduced IQ levels, behavioral problems, organ damage, anemia, convulsions, and death. Children face higher risks of health problems from lead exposure, but adults are affected as well. Pregnant women can experience damage to a fetus from lead. Exposure occurs through inhalation, ingestion, and dermatological contact. The U.S. Centers for Disease Control and Prevention (CDC) set 5 micrograms of lead per deciliter (μg/dL) of blood as a reference for public health actions, but there is no minimum level of exposure that is considered to be free of negative health effects. No level of exposure to lead is safe [19].

Lead-painted surfaces can produce a fine dust that is poisonous, especially to infants and children. This dust accumulates on floors under lead-painted windows and other building components. Toddlers crawl through this dust and ingest it through hand-to-mouth contact. This can also occur as children play outside in lead-contaminated soil. These hazards can be reduced or eliminated by following Lead Safe Work Practices to remove or encapsulate lead-based paint on a home's interior and exterior surfaces [20]. For soil contaminated with lead from paint chips or vehicle exhaust, that soil should be replaced, or barriers such as bushes should be planted to discourage children from playing in that soil.

Lead is also present in many household products, including slow cookers; lipstick and other cosmetics; house keys; hair dyes; faux leather purses, sandals, and wallets; and others [21].

Consumer education on this topic is necessary to inform the public about this issue.

7. Asbestos

The term asbestos refers to naturally occurring silicate minerals that are heat-resistant and fibrous. The fibers are soft and can be easily incorporated into building materials. Chrysotile, or white asbestos, was most commonly used in construction materials. Asbestos is found in older homes. It was used as insulation on heating systems and heating ducts. In some older homes, it actually covers entire boilers. Asbestos was also a component of joint compound, sheet goods that were used as fire barriers behind wood-burning stoves, roof sealants, floor and ceiling tile adhesives, gaskets, and automobile brake pads. Asbestos was also used in ironing board covers and potholders.

Like lead, humans have used asbestos for thousands of years. In ancient Egypt, pharaohs were embalmed in asbestos cloth. Ancient Roman aristocrats used asbestos tablecloths and napkins. After these items were used, they were placed in fires to clean them. The ancient Roman philosopher, Pliny the Elder, wrote about asbestos-caused lung disease and how asbestos mining slaves suffered from lung disease and made crude respirators to protect themselves [22].

Asbestos exposure occurs when asbestos fibers become friable, or airborne, and are inhaled. These fibers are microscopic and cannot be seen. This makes it possible for someone to inhale a large amount of fibers without knowing it. The fibers become embedded deep within the lungs, and the body cannot expel them. Exposure causes asbestosis, a type of lung cancer, and mesothelioma, which is a cancer of the mesothelial lining of the lungs. These diseases begin to show symptoms 10–40 years after initial exposure to asbestos.

Asbestos abatement is not a do-it-yourself activity. Its removal and encapsulation are regulated in the USA and must be performed by certified abatement contractors. These contractors seal off the work area where asbestos will be removed and wear disposable full-body protective suits and full head protection with respirators.

8. The child breathing zone

Children face higher risks than adults do from being exposed to environmental toxicants and from health problems caused by such exposure [23]. This is because children breathe larger amounts of air per body size when compared to adults. Sucking on hands and toys that have accumulated pollutants adds to these risks [24]. Another source of VOC exposure to infants is those that are emitted from crib mattresses and crib mattress covers [25].

The fact that 80% of children's alveoli are formed postnatally, and changes in the lung continue through adolescence, make children more vulnerable to developing health problems from air pollutants [26]. During the early postneonatal period, developing lungs are very susceptible to pollutants; and the immature immune, pulmonary, and nervous systems of children can be damaged by environmental pollutants, including routinely applied residential pesticides.

Young and older infants and young children breathe through their mouths than adults do. This difference in breathing patterns is likely to increase a child's risk of exposure to respirable particulates [27]. This risk is lower for adults whose breathing through their noses causes air to become filtered as they breathe it through the upper respiratory airway [27].

Toddlers crawl on the floor and young children walk, run, and play on the floor. These factors cause the breathing zone of children to be much lower (up to 3 feet from the floor). This zone is known as the child breathing zone (CBZ) [28]. Walking-induced turbulence in a room causes resuspension of respirable particulates, and shorter people are exposed to more resuspended particulates than taller people. IAQ can be significantly worse in the CBZ than in the adult breathing zone (ABZ), and the assumption of uniform pollutant concentration in indoor environments can be an erroneous assumption of breathing concentration risk. Although there is an increasing awareness that children are vulnerable to poor IAQ in the scientific community, there is very limited research with a focus on IAQ in the CBZ. There is no current IAQ management system that specifically focuses on improving IAQ in the CBZ.

9. Conclusion

The most effective strategy for controlling indoor air pollution is to control the problem at its source. Ventilation is also important, especially in the case of moisture. Expel moisture to the outside through exhaust fans that are vented to the outdoors. In the case of combustion pollutants, regular servicing of heating systems and other appliances that are combustion based is necessary. Radon gas is a radioactive human carcinogen that is colorless, tasteless, and odorless. This pollutant can be controlled through mitigation in existing homes and with radon-resistant construction techniques in new homes. Exposure to some VOCs, which are present in building materials, paints, strippers, and other substances, can be hazardous to human health. Adequate ventilation should be provided when using these materials. Low- or no-VOC emitting products are now available and should be considered as safer alternatives. Lead and asbestos are present in older homes and apartments and pose considerable health risks to humans. Only trained professionals should perform abatement or encapsulation of both materials. Children are at a higher risk of health problems from pollutant exposure, especially because air in the child breathing zone is more polluted than it is in the adult breathing zone. Awareness of these issues is a critical first step in improving air quality in homes and apartments.

Author details

Joseph Laquatra
Cornell University, Ithaca, New York, USA

*Address all correspondence to: jl27@cornell.edu

References

[1] Cincinelli A, Martellini T. Indoor air quality and health. International Journal of Environmental Resources and Public Health. 2017;**14**(11):1286. DOI: 10.3390/ijerph14111286

[2] Leech JA, Smith-Doiron M. Exposure time and place: Do COPD patients differ from the general population? Journal of Exposure Science & Environmental Epidemiology. 2016;**16**(3):238-241. DOI: 10.1038/sj.jea.7500452

[3] Coelho C, Steers M, Lutzler L, Schriver-Mazzuoli. Indoor air pollution in old people's homes related to some health problems: A survey study. Indoor Air. 2005;**15**:267-274. DOI: 10.1111/j.1600-0668.2005.00371.x

[4] Franklin PJ. Indoor air quality and respiratory health of children. Paediatric Respiratory Reviews. 2007;**8**(4):281-286. DOI: 10.1016/j.prrv.2007.08.007

[5] Ponessa J. Indoor air quality. In: Carswell AT, editor. The Encyclopedia of Housing. Thousand Oaks, CA: SAGE Publications, Inc; 2012. pp. 392-396. DOI: 10.4135/9781452218380.n137

[6] Arlian LG. Water balance and humidity requirements of house dust mites. Experimental and Applied Acarology. 1992;**16**(1-2):15-35. DOI: 10.1007/BF01201490

[7] Taylor T, Mendon V, Zhao M. Cost-Effectiveness of Heat Recovery Ventilation. 2015. Available from: https://www.energycodes.gov/sites/default/files/documents/iecc2018_R-3_analysis_final.pdf [Accessed: April 19, 2018]

[8] Tech Note: Whole-House Mechanical Ventilation Code: Safety and Performance Considerations. 2013. Available from: http://www.homeinnovation.com/~/media/Files/Reports/TechNote_WH_Ventilation_10252013.pdf [Accessed: April 19, 2018]

[9] Shin SH, Jo WK. Temporal characteristics of volatile organic compounds in newly-constructed residential buildings. Environmental Engineering Research. 2013;**18**(3):169-176

[10] U.S. Environmental Protection Agency (EPA). Indoor Air Quality. 2017. Available from: https://www.epa.gov/indoor-air-quality-iaq/technical-overview-volatile-organic-compounds [Accessed: April 22, 2018]

[11] Porta M, Last JM. A Dictionary of Public Health. 2nd ed. Oxford: Oxford University Press; 2018. DOI: 10.1093/acref/9780191844386.001.0001

[12] U.S. Geological Survey. The Geology of Radon. 1995. Available from: https://certmapper.cr.usgs.gov/data/PubArchives/radon/georadon/3.html [Accessed: August 21, 2018]

[13] U.S. Environmental Protection Agency (EPA). Health Risk of Radon. 2018. Available from: https://www.epa.gov/radon/health-risk-radon [Accessed: April 22, 2018]

[14] U.S. Environmental Protection Agency (EPA). A Citizen's Guide to Radon. 2016. Available from: https://www.epa.gov/sites/production/files/2016-12/documents/2016_a_citizens_guide_to_radon.pdf [Accessed: September 25, 2018]

[15] U.S. Environmental Protection Agency (EPA). Radon-Resistant Construction Basics and Techniques. 2018. Available from: https://www.epa.gov/radon/radon-resistant-construction-basics-and-techniques [Accessed: April 22, 2018]

[16] Hampson NB. U.S. mortality due to carbon monoxide poisoning, 1999-2014.

Accidental and intentional deaths. Annals of the American Thoracic Society. 2016;**13**(10):1768-1774

[17] Chauhan AS, Bhadauria R, Singh AK, Lodhi SS, Chaturvedi DK, Tomar VS. Determination of lead and cadmium in cosmetic products. Journal of Chemical and Pharmaceutical Research. 2010;**2**:92-97

[18] Jacobs DE, Clickner RP, Zhou JY, Viet SM, Marker DA, Rogers JW, Zeldin DC, Broene P, Friedman W. The prevalence of lead-based paint hazards in U.S. housing. Environmental Health Perspectives. 2002;**110**:A599-A606. Available from: http://ehpnet1.niehs.nih.gov/docs/2002/110pA599-A606jacobs/abstract.html [Accessed: May 8, 2018]

[19] American Academy of Pediatrics. AAP Commends CDC for Recognizing that for Children, There is no Safe Level of Lead Exposure. 2012. Available from: https://www.aap.org/en-us/about-the-aap/aap-press-room/Pages/AAP-Statement-CDC-Revised-Lead-Exposure-Guidelines.aspx [Accessed: May 21, 2013]

[20] U.S. Department of Housing and Urban Development (HUD). Lead Paint Safety. A Field Guide for Painting, Home Maintenance, and Renovation Work. 2001. Available from: https://www.hud.gov/sites/documents/DOC_11878.PDF [Accessed: May 8, 2018]

[21] Laquatra J. Lead in household products. In: Snedeker SM, editor. Toxicants in Plastics and Paper: Exposure and Health Risks to Consumers from Household Products and Food Packaging, Molecular And Integrative Technology. London: Springer-Verlag; 2014. pp. 231-243. DOI: 10.1007/978-1-4471-6500-2_9

[22] Povtak T. What Is Asbestos? Available from: https://www.asbestos.com/asbestos/ [Accessed: May 17, 2018]

[23] Canha N, Mandin C, Ramalho O, Wyart G, Ribéron J, Dassonville C, et al. Assessment of ventilation and indoor air pollutants in nursery and elementary schools in France. Indoor Air. 2016;**26**(3):350-365

[24] Klepeis NE, Nelson WC, Ott WR, Robinson JP, Tsang AM, Switzer P. The National Human Activity Pattern Survey (NHAPS): A resource for assessing exposure to environmental pollutants. Journal of Exposure Analysis and Environmental Epidemiology. 2001;**11**(3):231-252

[25] Boor BE, Järnström H, Novoselac A, Xu Y. Infant exposure to emissions of volatile organic compounds from crib mattresses. Environmental Science & Technology. 2014;**8**(6):3541-3549

[26] Dietert RR, Etzel RA, Chen D, Halonen M, Holladay SD, Jarabek AM, Landreth K, Peden DB, Pinkerton K, Smialowicz RJ, Zoetis T. Workshop to identify critical windows of exposure for children's health: Immune and respiratory systems work group summary. Environmental Health Perspectives. 2000;**108**(S3):483-490

[27] Flynn E, Matz P, Woolf A, Wright R. Indoor Air Pollutants Affecting Child Health. American College of Medical Toxicology. 2000. Available from : https://www.allergycosmos.co.uk/wp-content/uploads/2010/02/IndoorAirPolution.pdf [Accessed: May 18, 2018]

[28] Tripathii E, Laquatra J. Managing indoor air quality in the child breathing zone: Risk analysis and mitigation. Journal of Architectural Engineering. 2018;**24**(1):04018002-1-9. DOI: 10.1061/(ASCE)AE.1943-5568.0000300

Chapter 3

Indoor Air Quality Monitoring for Enhanced Healthy Buildings

Gonçalo Marques and Rui Pitarma

Abstract

Since most people spend 90% of their time indoors, the indoor environment has a determining influence on human health. In many instances, the air quality parameters are very different from those defined as healthy values. Using real-time monitoring, occupants or the building manager can decide and control behaviors and interventions to improve indoor air quality. The historical database is also useful for assisting doctors to support the medical diagnosis. The continuous technological advancements notably, as regards, networking, sensors, and embedded devices have made it possible to monitor and provide assistance to people in their homes. Smart objects with great capabilities for sensing and connecting could revolutionize the way we are monitoring our environment. This chapter consists of a general overview of several real-time monitoring systems developed and published by the authors. In this chapter, the authors present several new open-source and cost-effective systems that had been developed for monitoring environmental parameters, always with the aim of improving indoor air quality for enhanced healthy buildings.

Keywords: indoor air quality (IAQ), healthy buildings, occupational health, real-time monitoring, Internet of Things (IoT), ambient assisted living (AAL), enhanced living environments

1. Introduction

Indoor environments could be characterized by several pollutant sources. Environmental Protection Agency (EPA) is responsible for environmental air quality index regulation in the United States. This independent agency deliberates that indoor levels of contaminants can be up to 100 times greater than outdoor contaminant level and positioned poor air quality as one of the top five environmental dangers to the community well-being [1]. Thus, indoor air quality (IAQ) is recognized as an essential factor to be controlled for the occupants' health and comfort. Increase in the IAQ is critical as people typically spend more than 90% of their time in indoor environments. The problem of inadequate IAQ is of utmost importance affecting particularly severe form the poorest people in the world who are most vulnerable, presenting itself as a severe problem for world health such as tobacco use, alcoholism or the problem of sexually transmitted diseases [2].

In 1983, the World Health Organization (WHO) used the term "sick building syndrome" (SBS) to the clinical features that we might discover in building residents as a consequence of the poor IAQ [3]. Numerous statements have reported the influence of IAQ in the etiopathogenesis of various generic signs and medical results that illustrate SBS. The scientific representation of this pattern is widespread as it can engage the skin (with xerosis, pruritus), the upper and lower breathing tract (such as, dysphonia, dry cough and asthma), the eyes (ocular pruritus), and the nervous system (for example, headache and difficulty in concentration) [4, 5]. Furthermore, besides the symptoms of this disease, there are syndromes, which could be connected with indoor environments, i.e., Legionnaire's disease, extrinsic allergic alveolitis, asthma, and atopic dermatitis [4, 5]. For example, regarding atopic dermatitis, it is a chronic and inflammatory skin disorder and one of the most usual allergic syndromes in infants. Its occurrence is rising and, while it is related to hereditary influences, there is a considerable suggestion of responsibility for environmental factors, namely indoor air pollutants. This is mainly significant in industrialized nations, where youngsters apply most of their time inside buildings [6]. Including the air contaminants, the volatile organic compounds are connected to the exacerbation of atopic dermatitis, which remain the utmost deliberated usual pollutants of indoor air [5]. Universally acknowledged, in atopic dermatitis, indoor air contaminants could provoke oxidative stress, leading to skin barrier dysfunction or immune dysregulation [6]. Thus, the signs and syndromes related to the "sick buildings" are a problem with emergent significance in public health and have likewise been associated with lower productivity and greater absenteeism. The etiology of the SBS and the building associated disorders might incorporate chemical pollutants (both from outdoor and indoor sources), biological agents, emotional issues, electromagnetic radiation, the deficiency of sunlight, humidity, poor acoustics, deficient ergonomics, and bad ventilation [5]. Although the importance of indoor air quality for public health still exists, there is a lack of interest in the new scientific methods to improve indoor air quality in developed countries [7].

Ventilation is used in buildings to create thermally comfortable environments with acceptable IAQ by regulating indoor air parameters, such as air temperature, relative humidity, airspeed, and chemical species concentrations in the air [8]. An IAQ evaluation system provides an important way to find and enhance the indoor environmental quality. Local and distributed valuation of chemical concentrations is substantial not only for security (gas spills recognition, pollution supervising) and well-being applications but also for efficient temperature regulation, ventilation and air conditioning (HVAC) system for energy efficiency [9]. IAQ monitoring offers an uninterrupted stream of data for centralized regulation of building automation procedures, and delivers a solution for enhanced build management [10]. Real-time supervision of the IAQ is assumed as an essential tool of extreme importance to plan interventions for enhanced occupational health.

In recent past, numerous systems have been created on behalf of environmental supervision, constantly beside the intention to increase the IAQ [11]. The accessibility of cost-effective, energy efficient, and small-scale embedded computers, radios, sensors, and actuators, regularly incorporated on a unique chip, has been conducted for the incorporation of wireless communications to cooperate with the material world for IAQ supervision and enhanced living environments [12].

In this chapter, the authors present several new open-source and cost-effective systems that had been developed for monitoring environmental parameters, always with the aim of improving IAQ for enhanced healthy buildings. The chapter is structured as follows: besides the Introduction (Section 1), Section 2 introduces IoT and AAL themes, and Section 3 is concerned with presenting several IAQ systems for enhanced living environments, and Section 3.2 demonstrates a comparison between the proposed systems, and the conclusions are presented in Section 3.5.

2. IoT and AAL for the enhanced indoor air quality

Ambient assisted living (AAL) is closely related to the necessity of pervasive healthcare supervision, and the main aim is to contribute to the pervasion of the independence and well-being for older adults using Information and Communication Technologies (ICTs) [13].

Nowadays, there are numerous AAL solutions that can be found in literature that incorporate a large number of different types of sensors for biological supervision. These solutions typically incorporate wireless communication technologies for data sharing and collection such as *ZigBee, Bluetooth, Ethernet*, and *Wi-Fi*.

At the second half of this century, 20% of the humankind will be of age 60 or above [14], which is linked with several complex problems for public health. First of all, this will provoke an increase in disorders, healthcare budget, and the scarcity of caretakers, which will conduct to a giant social impact. Another import argument is that people typically choose to remain in their homes even paying the cost of the nursing care [15], which indicate the research of AAL solutions architectures as unquestionably a subject of extraordinary significance taking into account the humankind aging.

AAL researches are planned to encounter the requirements of the elderly population to preserve their independence as long as conceivable. On the one hand, improvements in telecommunications, sensors, and embedded processors conducted to the delivery of real-time supervising and personalized healthcare solutions to entities, which are able to be currently used in their habitats. On the other hand, these incessant scientific developments create the elaboration of smart cyber-physical systems for enhanced living environments and occupational health. Although there is a portion of issues in the creation of an effective AAL ecosystem such as data architecture, interface design, human-computer communication, ergonomics, usability and availability [16], there are also collective and moral difficulties as the recognition by the older people and the privacy and confidentiality that would stand as a prerequisite of the entirely AAL solutions. Indeed, it is likewise crucial to guarantee that technology does not substitute the human care but instead must be an extraordinary compliment.

Internet of Things (IoT) stands as a standard where things are linked to the Internet and incorporate data collection capabilities. The basic idea of the IoT is the pervasive presence of a variety of objects with interaction and cooperation capabilities among them to reach a common objective [17–19]. It is anticipated that the IoT will provoke a considerable effect on numerous characteristics of daily life and this paradigm will be incorporated in several purposes such as domotics, assisted living, e-health and is likewise a perfect emergent knowledge to offer novel evolving data and computational resources on behalf of generating groundbreaking software applications [20]. IoT architectures should incorporate wireless communication technologies. Nowadays, several wireless communication technologies are available such as bluetooth-based technologies, Wi-Fi-based technologies, near-field communication (NFC)-based technologies, and GSM-based technologies.

IoT solutions must stand pervasive, be context aware, and allow environment intelligence skills that are directly connected to AAL. IoT is an appropriate method to construct well-being solutions. Scientific developments turn possible to create novel and innovative instruments to empower real-time healthcare supervising solutions for decision-making in the management of several syndromes.

Nowadays, several IoT architectures had been implemented for clinical monitoring that claim IoT as a reliable platform to develop personalized healthcare systems. Due to Bluetooth technology, the use of wearables for data collection and smartphones for data transmission is now possible to provide physiological parameter supervision [20]. In 2009, several research initiatives for remote healthcare was been developed using IoT. Furthermore, IoT can increase the knowledge of data collection, which support IoT solutions in the medical area [21]. On behalf of the potential of the IoT concept for wellbeing solutions nowadays several challenges to be overcome still subsist.

The influence and impact of the IoT in the today's market is not clearly known, as well as the acceptance of pervasive and ubiquitous IoT products. Although the scientific advancements that turn IoT healthcare systems currently are feasible, the timing might be too early [22].

The "smart city" conception has lately presented as a tactical strategy to face contemporary municipal manufacture features in a mutual framework and, in specific, to focus the significance of ICT in the previous 20 years for increasing the economical profile of a city as suggested by [23]. Currently, cities have fascinating challenges and complications to gather socioeconomic progress and quality of life intents. The "smart cities" are related to react to these problems [24]. The smart city is also straightly connected to an emergent approach to decrease the difficulties produced by the urban population progress and quick urbanization [25]. The highest significant challenge in smart cities is the interoperability of the diverse technologies. IoT might be able to offer the interoperability to develop an integrated urban-scale ICT architecture for smart cities [26]. The smart city execution will produce effects at diverse stages such as effects on science, effects on technology and competitiveness, and effects on culture; however, this will likewise provoke ethical concerns as the smart city requires to offer accurate data access as it becomes fundamental, once such data are accessible at a fine spatial scale where people can be recognized [27]. IoT has an important potential to build novel real-life solutions and services for the smart city background [28].

3. Indoor air quality monitoring systems

Several solutions have been developed to improve the occupational health, aiming to provide real-time monitoring of indoor environments for enhanced living environments and occupational health. These solutions could revolutionize the indoor environments contributing to enhanced healthy buildings and to decrease the SBS problem. Some systems developed by the authors are described below.

3.1 *iAQ system*

iAQ system [29] is an automatic low-cost indoor air quality monitoring wireless sensor network system, developed using *Arduino*, *XBee* modules, and microsensors. This solution can be accessed by the building supervisor to identify a diversity of factors as temperature, humidity, luminosity, carbon dioxide (CO_2), and carbon monoxide (CO), in real time. Other parameters can be analyzed for particular contaminants as other sensors might be added for data collection.

The *iAQ Sensor* is responsible for the environmental data collection and to transmit these data to the *iAQ Gateway*. The *iAQ Gateway* uses Web services to provide data transmission and storage in a *MySQL* database. The Web services was been developed in PHP (**Figure 1**).

The iAQ system can incorporate one or more iAQ Sensors (**Figure 2**). The *iAQ Sensors* not only collect data from the environment but also send these data to the *iAQ Gateway* and can be placed in different locations. The *iAQ Gateway*, which is

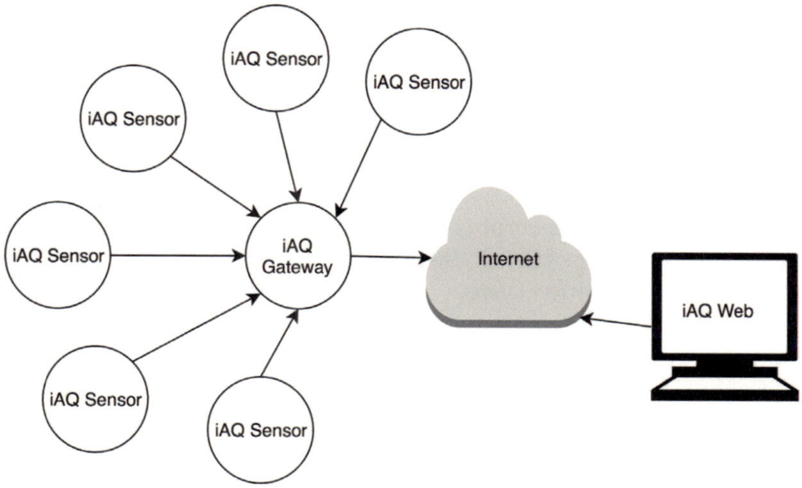

Figure 1.
iAQ WSN architecture, from [29].

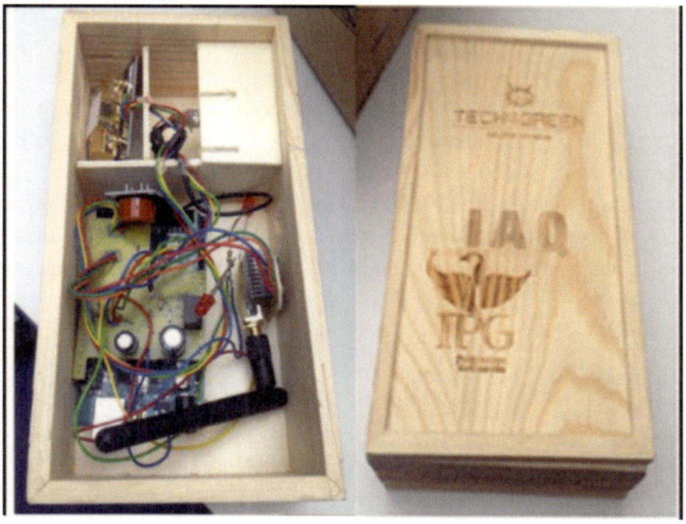

Figure 2.
iAQ Sensor hardware, from [29].

linked to the Internet through an *Arduino* Ethernet Shield, is responsible for the data storage.

The Web portal *iAQ Web* was been developed in PHP and supports authentication. This Web portal is responsible for the availability of the data to the end user. Accessing the *iAQ Web* is possible to analyze the environmental quality. The data can be analyzed as numeric values or in chart form. The *iAQ Web* is prepared with a notification manager that notifies the user in case a particular parameter overdoes the maximum value. This portal additionally acknowledges the user to retain the parameters' history. Offering a history of changes, this system provides an evaluation platform to accurately analyze the IAQ behavior. Furthermore, it is possible to take actions in the environment to increase the air quality in the building in real time.

The wireless communication features are created with the *ZigBee* networking protocol. Several *XBee* modules are used to implement the *IEEE 802.15.4* radio standard [30]. This standard identifies the physical and medium access control layers for low data-rate personal networks. *ZigBee* is a cost-effective, energy efficient, support mesh networks standard and was develop upon 802.15.4. Radio waves are transmitted from *iAQ Sensor* to the base station *iAQ Gateway*.

3.2 *iAQ mobile system*

Currently, smartphones incorporate high processing specifications aside from a diversity of sensors appropriated to the research and development of AAL systems. Sensors such as global position system (*GPS*), bluetooth low energy (*BLE*), camera, microphone, luminosity, accelerometer, gyroscope, and near-field communication (NFC).

As for the importance of the smartphone's role in human life, *iAQ* solution has been updated with an *Android* application [31, 32]. This mobile application was designed to provide quick and easy access to *iAQ system* to allow the end user to keep all the relevant information of *iAQ system* in your pocket.

This application provides data authentication and protection mechanisms for information visualizations and allows one to view system data in detail and receive notifications when any of the values exceed normal values.

This mobile application was developed for the android mobile operating system. The *integrated development environment* (IDE) *Android Studio* was used to build the application. The minimum requirement is the *application programming interface* (API) 15: *Android* 4.0.3 Ice Cream Sandwich. According to the IDE, this mobile app is compatible with 96.2% of active devices in the *Google play store* (information collected on January 22, 2016).

Figure 3 represents the mobile application features. The left image represents the login screen of the application that guarantees the authentication before data access. The right image represents the ability to select one of the monitored

Figure 3.
Android app, from [33].

locations that correspond to a specific *iAQ Sensor* node. The user can access to the current humidity, temperature, carbon dioxide, carbon monoxide, and light values.

Using the mobile app, the user can likewise rapidly access to the alerts generated when the monitored parameters exceed the minimum or the maximum values.

3.3 *iAQ IoT system*

iAQ solution has also been updated to adopt an IoT architecture using the *ESP8266* and be a fully wireless solution for IAQ. *iAQ IoT Gateway* [33] has replaced the *Arduino* by a *Wemos Mini D1 (Wemos Electronics)* as a processing unit. The processing unit is a miniaturized *Wi-Fi* board based on *ESP-8266EX*. This board incorporates 11digital input pins and 1 digital output pin, and 1 analogue input pin. The interface used for programming and power is micro-USB. *Wemos Mini D1* can be programmed using the *Arduino* IDE and incorporate 32 bits CPU with an 80/160 MHz clock speed, which works at 3.3 V, 4 Mb flash, and has 34.2 × 25.6 mm size and a weight of 10 g.

A majority of the IAQ supervision solutions currently available on the market are especially expensive and only permit the collection of arbitrary values from the environment. *iAQ* IoT is an IAQ solution developed on top of the IoT concept that integrates in its assembly Arduino, *ESP8266*, and *XBee* technologies for data processing and transmission and sensors for data collection.

iAQ IoT Gateway incorporates only wireless communication technologies to interact with the nodes as well as for Internet accessibility. It collects data from *iAQ Sensor* using an *XBee* module and then uses Wi-Fi to provide data storage to a *MySQL* database using Web services. The schematic and connections used in *iAQ Gateway* are described in **Figure 4**.

3.4 *iAQ Wi-Fi system*

iAQ Wi-Fi [34] is a real-time indoor air quality monitoring solution that is capable of measuring temperature, humidity, PM10, CO_2, and luminosity in real time. This solution is based on the IoT concept and is fully wireless. The access to the Internet in order to provide data storage of the monitored parameters is developed using the *ESP8266* chip, which implements the *IEEE 802.11 b/g/n* networking protocol and supports radio transmission within the 2.4 GHz band.

This solution incorporates open-source technologies, using an *Arduino UNO* as a microcontroller as processing unit and an *ESP8266* module as the communication unit. The monitored data are uploaded to the *ThingSpeak* platform. *ThingSpeak* is an

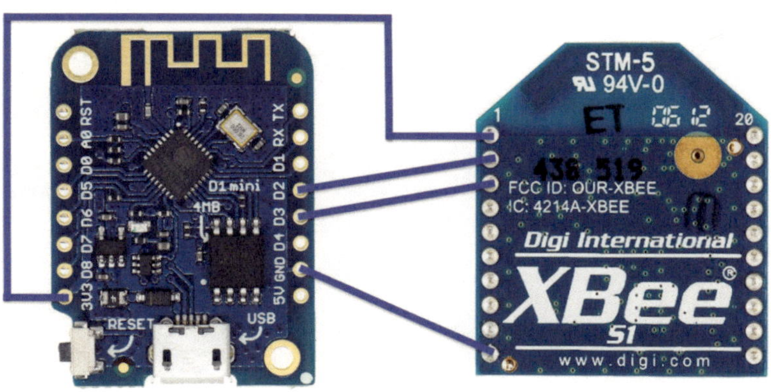

Figure 4.
iAQ IoT Gateway, from [33].

open-source IoT platform that offers *APIs* for storing and retrieving data from sensors and devices using *HTTP* [12]. *iAQ Wi-Fi* prototype is represented in **Figure 5**.

The end user can access the data from the Web page provided by *ThingSpeak* platform or can use the smartphone app developed in *SWIFT*, an open-source programming language with *XCODE* IDE created for iOS operating system. *iAQ Wi-Fi* system architecture is based on IoT. **Figure 6** represents the system architecture used by the authors.

The *Arduino UNO* incorporates sensors that are responsible for the data collection and send that information to the *ESP8266* by serial communication. The

Figure 5.
iAQ Wi-Fi prototype, from [34].

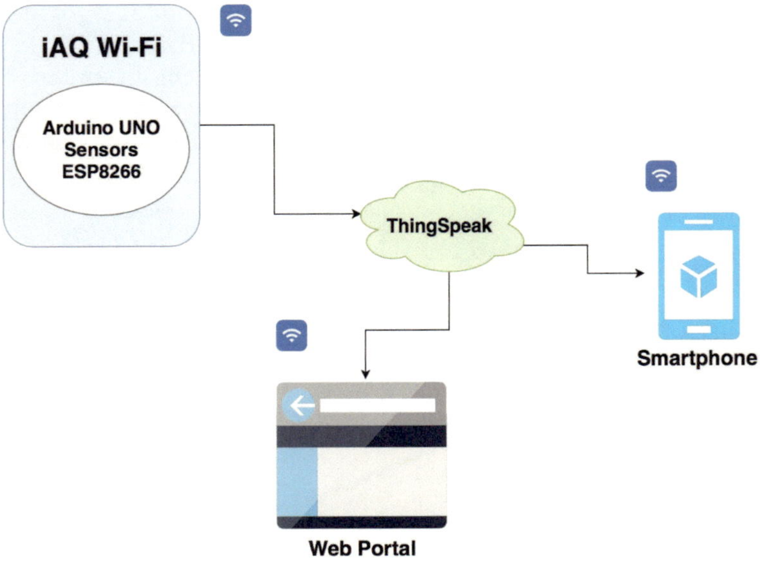

Figure 6.
iAQ Wi-Fi system architecture, from [34].

ESP8266 is connected to the Internet using Wi-Fi and is responsible for uploading the data received to the *ThingSpeak* platform.

The mobile application is denominated by *iAQ Wi-Fi Mobile* (**Figure** 7), and is developed using *XCODE* IDE and *SWIFT* programming language. The *iAQ Wi-Fi Mobile* has the iOS 7 as the minimum requirement. The mobile application provides authentication to authorized users. The end user after login can analyze the history-monitored data in numerical or graphical representation.

3.5 *iAirC system*

iAirC is a solution for carbon dioxide (CO_2) real-time monitoring based on IoT architecture. To have a low-cost system, only one type of indoor air pollutant was chosen [35].

In one hand, when the CO_2 level extends 7–10%, an individual can lose consciousness within minutes and might stand at risk of death. On the other hand, a low intensity of CO_2 stands inoffensive to humans. It is well known that CO_2 levels are linked with dizziness and sleepiness leading to low productivity at work [36]. Therefore, it is significant to provide a real-time CO_2 supervision and develop a notification system for enhanced living environments. The intensity of CO_2—the main greenhouse gas—is steadily increased to 400 ppm (ppm), reaching new records every year since they began to be produced in 1984 [37].

CO_2 was chosen because it is easy to measure, and it is produced in quantity from multiple sources (by people and combustion equipment). Consequently, it should be assumed as an indicator of other contaminants, and consequently of IAQ in common.

The *iAirC* solution incorporates a prototype for environmental data collection and a mobile application for data access and supervision. This solution use Wi-Fi for Internet access, which conduct to a diversity of advantages such as modularity, scalability, low-cost and easy installation. The data are stored in the *ThingSpeak*

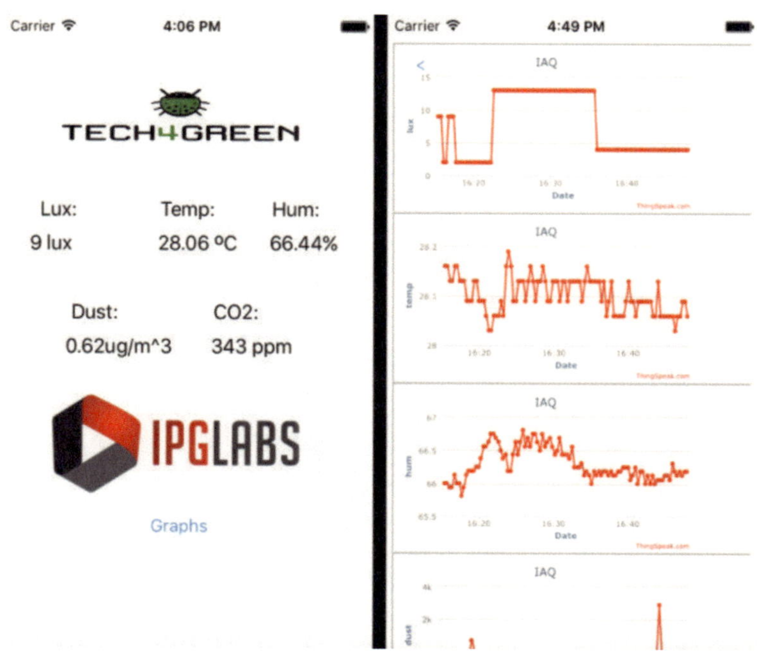

Figure 7.
iAQ Wi-Fi mobile app, from [34].

cloud platform and then can be consulted using the mobile app or *ThingSpeak* Web portal.

iAirC consists of two components, an *ESP8266 Thing Dev (Sparkfun) microcontroller* and an MHZ-19 carbon dioxide sensor developed by *Winsensor* (**Figure 8**).

The *ESP8266 Sparkfun* microcontroller incorporates integrated *Wi-Fi* features and is used mutually for data processing and communication. The *iAirC* is a low-cost, reliable system that can be easily configured and installed by the average user. For this, *iAirC* incorporates a low cost but very reliable carbon dioxide sensor and a microcontroller with native *Wi-Fi* support.

The mobile application is denominated by *iAirC Mobile* (**Figure 8**), and is developed using *XCODE* IDE and *SWIFT* programming language [38]. Using the *iAirC Mobile,* the end user after authentication not only can access to real-time CO_2 levels but also to be notified when the IAQ is defective (**Figure 8**).

Ample physical evidence shows that CO_2 is the single most important climate-relevant greenhouse gas in Earth's atmosphere and high external charges mean that they naturally lead to higher indoor concentrations due to the contribution of the internal sources (human metabolism and combustion equipment) [39, 40]. It is imperative to control the concentration of CO_2 effectively and the authors believe that the first step is to monitor to perceive its variation in real time and to plan interventions for its reduction.

3.6 *iDust*

PM is related to numerous serious health problems. *iDust* is a real-time PM exposure monitoring system and decision-making tool for enhanced healthcare based on an IoT architecture [41]. It was developed using open-source technologies and low-cost sensors.

This architecture has been developed in order to provide an evaluation platform that can be acceded to by the building manager in order to analyze the PM behavior of the indoor environment in detail. Furthermore, the build manager can take action in real time in order to provide a safe and healthful place for the occupants.

Figure 8.
iAirC prototype (left); and iAirC Mobile (right), from [35].

PM exposure data can be exceptionally precious to provide support to a clinical examination by medical experts as the therapeutic panel could analyze the record of IAQ factors of the environment everywhere the patient resides and link this reports alongside his health problems. On the other hand, by supervising IAQ, it stands plausible to identify the air quality circumstances appropriately plus, if required, plan interventions to drop the PM exposure concentrations.

iDust stands as an ICT solution for real-time IAQ managing that allows the end user, as the building manager to analyze the PM exposure behavior. This system incorporates *WEMOS D1 mini* as a microcontroller and is developed using the *Arduino IDE*. The parameters are supervised using the *iDust* prototype, which is responsible for data collection. The monitored data are stored in a *SQL SERVER* database using Web services built in .NET. An authenticated user is able to access the IAQ information using the Web portal created in *ASP.NET*. The information collected is accessible in a dashboard in mutually numeric values or graph form. Likewise, the Web portal stores the history of the PM exposure behavior. The Web application incorporates a important notification manager that notifies the build manager when a particular parameter exceeds the maximum value. *iDust* is a cost-effective, consistent method, which can easily be parametrized and installed by the regular people. On behalf of this, the authors had selected a cost-effective but very reliable PM sensor and a microcontroller with built-in *Wi-Fi* communication technology. This architecture incorporates of two components: a microcontroller and a PMS5003 PM sensor (*Plantower*), which features scattering method to quantify the rate of particles suspended in the air with a diameter of 10 microns or less (≤PM10), 2.5 microns or less (≤PM2.5), and 1.0 microns or less (≤PM1.0) (**Figure 9**).

The Web application allows the build manager to save the parameters' history as is presented in **Figure 10**. Providing a history of changes turns possible to do an accurate and detailed analysis of the PM exposure behavior.

iDust system uses the *ESP8266* for both processing and Internet connectivity. The incorporation of the *ESP8266* has an additional significant functionality as it offers to the regular user an easy configuration of the *Wi-Fi* network. The *ESP8266* is by default a *Wi-Fi* client, although when a known *Wi-Fi* network is not available, or in case there are no wireless networks available, the *ESP8266* will turn to hotspot mode and will transmit a *Wi-Fi* network with an service set identifier (SSID) "IAQ-iDust." After that, the regular user could be connected to this Wi-Fi network to configure the *Wi-Fi* network to which the *iDust* is going to be connected. The regular user must introduce the SSID and password using a Web form as represented in **Figure 11**.

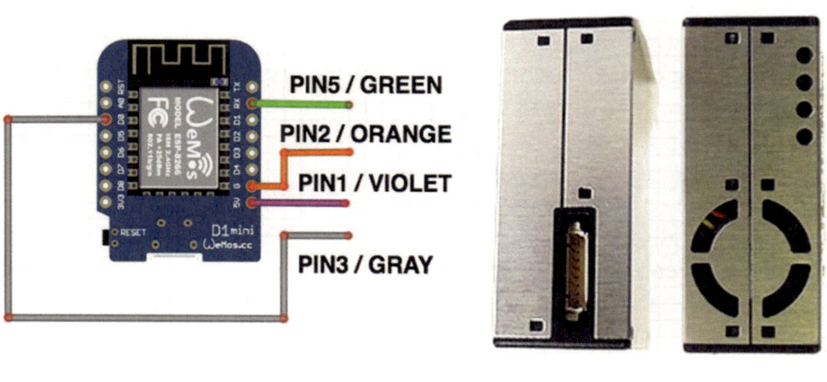

Figure 9.
iDust connection diagram, from [41].

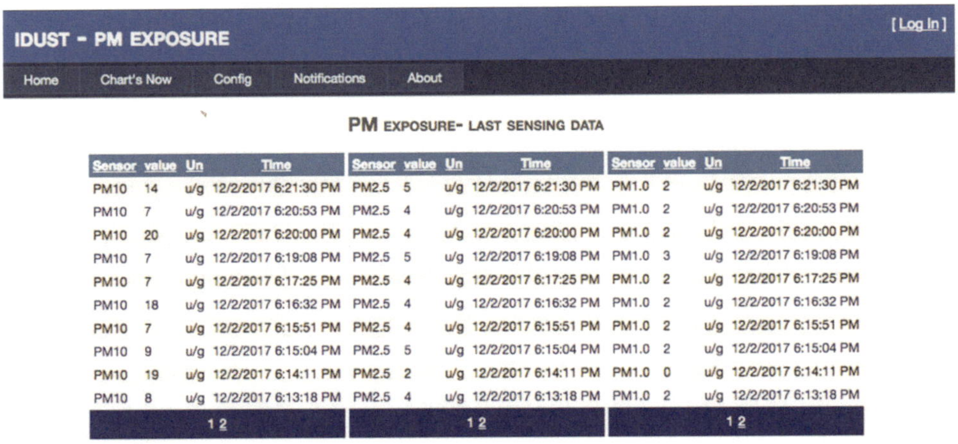

Sensor	value	Un	Time	Sensor	value	Un	Time	Sensor	value	Un	Time
PM10	14	u/g	12/2/2017 6:21:30 PM	PM2.5	5	u/g	12/2/2017 6:21:30 PM	PM1.0	2	u/g	12/2/2017 6:21:30 PM
PM10	7	u/g	12/2/2017 6:20:53 PM	PM2.5	4	u/g	12/2/2017 6:20:53 PM	PM1.0	2	u/g	12/2/2017 6:20:53 PM
PM10	20	u/g	12/2/2017 6:20:00 PM	PM2.5	4	u/g	12/2/2017 6:20:00 PM	PM1.0	2	u/g	12/2/2017 6:20:00 PM
PM10	7	u/g	12/2/2017 6:19:08 PM	PM2.5	5	u/g	12/2/2017 6:19:08 PM	PM1.0	3	u/g	12/2/2017 6:19:08 PM
PM10	7	u/g	12/2/2017 6:17:25 PM	PM2.5	4	u/g	12/2/2017 6:17:25 PM	PM1.0	2	u/g	12/2/2017 6:17:25 PM
PM10	18	u/g	12/2/2017 6:16:32 PM	PM2.5	4	u/g	12/2/2017 6:16:32 PM	PM1.0	2	u/g	12/2/2017 6:16:32 PM
PM10	7	u/g	12/2/2017 6:15:51 PM	PM2.5	4	u/g	12/2/2017 6:15:51 PM	PM1.0	2	u/g	12/2/2017 6:15:51 PM
PM10	9	u/g	12/2/2017 6:15:04 PM	PM2.5	5	u/g	12/2/2017 6:15:04 PM	PM1.0	2	u/g	12/2/2017 6:15:04 PM
PM10	19	u/g	12/2/2017 6:14:11 PM	PM2.5	2	u/g	12/2/2017 6:14:11 PM	PM1.0	0	u/g	12/2/2017 6:14:11 PM
PM10	8	u/g	12/2/2017 6:13:18 PM	PM2.5	4	u/g	12/2/2017 6:13:18 PM	PM1.0	2	u/g	12/2/2017 6:13:18 PM
1 2				1 2				1 2			

Figure 10.
iDust Web application, from [41].

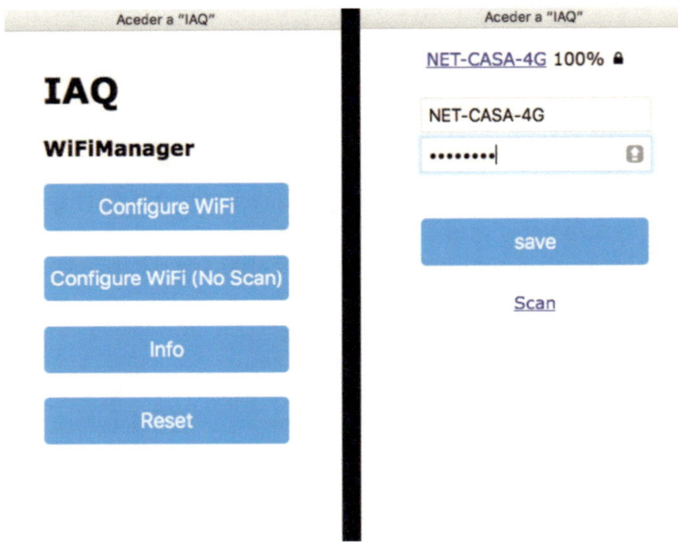

Figure 11.
iDust Wi-Fi configuration, from [41].

4. Discussion

Several solutions for IAQ supervision, which support open-source technologies for data processing, collection, and transmission that offers mobile computing architectures for real-time data accessibility, was presented in Section 3. Mainly, IAQ monitoring is a trending topic for which some other low-cost and open-source monitoring systems had been developed.

A summary of these studies is presented in **Table 1**.

In general, all the systems presented not only use cost-effective sensors and use open-source technologies, but also have notification systems that allow users to act in real time to significantly improve indoor air quality through the ventilation or deactivation of pollutant equipment. The presented solutions make a significant contribution compared to existing air quality monitoring systems due to its low cost of construction, installation, modularity, scalability, and easy access to monitoring data in real time through the Web and mobile applications. All the presented

	MCU	Sensors	Architecture	Low cost	Open-source	Connectivity	Data access	Data storage	User installation
iAQ	Arduino	Temperature, relative humidity luminosity, CO, CO_2	WSN	√	√	ZigBee, Ethernet	Web	MySQL	×
iAQ Mobile	Arduino	Temperature, relative humidity luminosity, CO, CO_2	WSN	√	√	ZigBee, Ethernet	Mobile	MySQL	×
iAQ IoT	Arduino, ESP8266	Temperature, relative humidity luminosity, CO, CO_2	WSN/IoT	√	√	ZigBee, Wi-Fi	Web, mobile	MySQL	×
iAQ Wi-Fi	Arduino, ESP8266	CO_2, particulate matter, temperature, relative humidity	IoT	√	√	Wi-Fi	Web, mobile	Cloud	√
iAirC	Sparkfun ESP8266	CO_2,	IoT	√	√	Wi-Fi	Mobile	Cloud	√
iDust	Wemos D1 Mini	Particulate matter	IoT	√	√	Wi-Fi	Web, mobile	SQL Server	√

MCU: microcontroller; √: apply; ×: not apply.

Table 1.
Summary of the presented systems for real-time indoor air quality monitoring.

solutions aim to offer the support to a medical examination by clinical professionals as the medical team might analyze the history of IAQ parameters collected from the environments where the patient lives and relate these records with his health complications.

An essential advantage of the use of *ZigBee* communication (*iAQ*, *iAQ Mobile* and *iAQ IoT*) is that we can have many *iAQ Sensors* collecting indoor air quality data, and only one *iAQ Gateway* must be connected to the Internet as *Zigbee* have an indoor RF line-of-sight range up to 50 m. It is essential for some scenarios, in this way, as it is no longer being required for *Wi-Fi* network coverage throughout all area of housing and it needs only an Internet connection at the location of *iAQ Gateway*.

Of about 56% of American adults are now smartphone holders [42]. In Netherlands, 70% of the regular people and over 90% of teenagers have a smartphone [43]. The usage of mobile phones represents on average 86 min per day (median 58 min) as proposed by [44]. People use the smartphone even when they are close to the computer [45]. Mobile computing offers the possibility to check the data and gather notifications anytime and anywhere. Real-time notifications provide a reliable method to maintain healthy indoor environment to increase the occupant's health, productivity, and well-being as the building manager can react at time. Consequently, mobile applications were been created in *iAQ Mobile*, *iAQ IoT*, *iAQ Wi-Fi*, and *iAirC*.

iAQ IoT system incorporates not only wireless communication technologies to interact between *iAQ Sensors* and *iAQ IoT Gateway* but also for Internet connection. Therefore, the implementation and installation cost is lower when compared with other solutions that use Ethernet to connect the gateway to the Internet. This solution could easily be designed to use only as many *iAQ Sensors* as needed due to their modularity and integrates the benefits of the *WSN* and IoT architectures.

iAQ Wi-Fi and *iAirC* use cloud service for data storage and data access. The use of cloud service has several advantages referring the cost and security of the storage data.

iAQ Wi-Fi, *iAirC*, and *iDust* have benefits both in easy installation and configuration, not only due to the use of wireless technology for communications, but also because they were developed to be compatible with all domestic house devices and not only for smart houses or high-tech houses. These systems are particularly useful for the analysis of IAQ. These functionalities create an easy product installation, which is directly related to IoT concept. The common IAQ supervising architectures should be installed by specialized professional, although the iAQ Wi-Fi, iAirC, and iDust solutions can be installed by the regular people using a gadget with Wi-Fi connectivity, which decrease the costs related to the installation.

Compared to other systems, *iAirC* and *iDust* systems incorporate only one sensor, which provides advantages both in ease of installation and configuration due to the use of wireless technology but also due to its small size. These systems use the *ESP8266* for both processing and Internet connectivity. This method not only delivers numerous benefits concerning the decrease of the system cost, but also increases the processing power as the *ESP8266* has an 80 MHZ CPU, while the *Arduino UNO* has a 16 MHZ CPU, for example.

5. Conclusions

This chapter has presented several solutions for indoor air quality monitoring and decision-making tools for enhanced healthcare. All the given solutions were developed using open-source technologies, cost-effective sensors, low price of construction, installation, modularity, scalability, and easy access to monitoring data. The results obtained by these solutions are auspicious, as this kind of systems might be used to provide a detailed stream of data that can be used by the building manager for correct maintenance to offer not only a safe but also a healthy environment for enhanced living environments.

On the one hand, the real-time monitoring is a significant method to support the clinical analysis by medical specialists as the therapeutic team could analyze the history of IAQ conditions of the environment where the patient resides and link these data with his health problems. On the other hand, by supervising IAQ, it is conceivable to identify the poor air quality situations appropriately and plan interventions for enhanced living environments.

The *WSN* architecture is appropriated to large buildings with no *Wi-Fi* networks available. However, the IoT architecture is appropriated to domestic homes as the majority provide Wi-Fi access points and also because the easy installation and configuration allows the user to start with a few devices and increase the number of them as he needs.

In the opinion of the authors, the future of air quality monitoring solutions focuses on the development of *Wi-Fi* systems that incorporate only one sensor. In this way, the user can not only create an ecosystem to suit them by monitoring the parameters he wants, but can also make the systems more cost-effective and easier to install.

As a future work, the proposed solutions should plan software and hardware improvements to fit specific cases such as hospitals, schools, and industry. It is also essential to create secure methods for data sharing between the medical team in order to support clinical diagnostics. The authors believe that in the future, systems like the presented ones will be used as an integral part of the daily human routine in order to provide safe and productive living environments.

Acknowledgements

The financial support from the Research Unit for Inland Development of the Polytechnic Institute of Guarda is acknowledged.

Conflict of interest

The authors declare no conflict of interest.

Author details

Gonçalo Marques and Rui Pitarma*
Unit for Inland Development, Polytechnic Institute of Guarda, Guarda, Portugal

*Address all correspondence to: rpitarma@ipg.pt

References

[1] Seguel JM, Merrill R, Seguel D, Campagna AC. Indoor air quality. American Journal of Lifestyle Medicine. 2016;**11**(4):284-2895

[2] Bruce N, Perez-Padilla R, Albalak R. Indoor air pollution in developing countries: A major environmental and public health challenge. Bulletin of the World Health Organization. 2000; **78**(9):1078-1092

[3] Jafari MJ et al. Association of sick building syndrome with indoor air parameters. Tanaffos. 2015;**14**(1):55

[4] Capristo C, Romei I, Boner AL. Environmental prevention in atopic eczema dermatitis syndrome (AEDS) and asthma: Avoidance of indoor allergens. Allergy. 2004;**59**(s78): 53-60

[5] Joshi S. The sick building syndrome. Indian Journal of Occupational and Environmental Medicine. 2008;**12**(2):61

[6] Ahn K. The role of air pollutants in atopic dermatitis. Journal of Allergy and Clinical Immunology. 2014;**134**(5): 993-999

[7] Sundell J. On the history of indoor air quality and health. Indoor Air. 2004;**14** (s7):51-58

[8] Pitarma R, Lourenço M, Ramos J. Improving occupational health by modelling indoor pollutant distribution. Facilities. 2016;**34**(5/6):289-301

[9] De Vito S et al. Cooperative 3D air quality assessment with wireless chemical sensing networks. Procedia Engineering. 2011;**25**:84-87

[10] Preethichandra DMG. Design of a Smart Indoor Air Quality Monitoring Wireless Sensor Network for Assisted Living; 2013. pp. 1306-1310

[11] Yu T-C, Lin C-C. An intelligent wireless sensing and control system to improve indoor air quality: Monitoring, prediction, and preaction. International Journal of Distributed Sensor Networks. 2015;**11**(8):140978

[12] Al-Haija QA, Al-Qadeeb H, Al-Lwaimi A. Case study: Monitoring of AIR quality in King Faisal University using a microcontroller and WSN. Procedia Computer Science. 2013;**21**: 517-521

[13] Universal Open Platform and Reference Specification for Ambient Assisted Living. http://www.universaal.org/

[14] UN. Worldpopulationageing: 1950–2050; 2001. pp. 11-13

[15] Centers for Disease Control and Prevention. The state of aging and health in America 2007. N. A. on an Aging Society; 2007. Available: https://www.cdc.gov/aging/pdf/saha_2007.pdf

[16] Koleva P, Tonchev K, Balabanov G, Manolova A, Poulkov V. Challenges in designing and implementation of an effective ambient assisted living system. In: Telecommunication in Modern Satellite, Cable and Broadcasting Services (TELSIKS), 2015 12th International Conference on; 2015. pp. 305-308

[17] Giusto D, editor. The Internet of Things: 20th Tyrrhenian Workshop on Digital Communications. New York: Springer; 2010

[18] Marques G, Pitarma R. Monitoring and Control of the Indoor Environment. In: 2017 12th Iberian Conference on Information Systems and Technologies (CISTI); 2017. pp. 1-6

[19] Marques G, Pitarma R. Monitoring energy consumption system to improve

energy efficiency. In: Rocha Á, Correia AM, Adeli H, Reis LP, Costanzo S, editors. Recent Advances in Information Systems and Technologies. Vol. 570. Cham: Springer International Publishing; 2017. pp. 3-11

[20] Gubbi J, Buyya R, Marusic S, Palaniswami M. Internet of Things (IoT): A vision, architectural elements, and future directions. Future Generation Computer Systems. 2013;**29**(7):1645-1660

[21] Luo J, Chen Y, Tang K, Luo J. Remote monitoring information system and its applications based on the Internet of Things. In: BioMedical Information Engineering, 2009. FBIE 2009. International Conference on Future. 2009. pp. 482-485

[22] Swan M. Sensor Mania! The Internet of Things, Wearable Computing, Objective Metrics, and the Quantified Self 2.0. Journal of Sensor and Actuator Networks. 2012;**1**(3):217-253

[23] Caragliu A, Del Bo C, Nijkamp P. Smart Cities in Europe. Journal of Urban Technology. 2011;**18**(2):65-82

[24] Schaffers H, Komninos N, Pallot M, Trousse B, Nilsson M, Oliveira A. Smart cities and the future internet: Towards cooperation frameworks for open innovation. In: Domingue J, Galis A, Gavras A, Zahariadis T, Lambert D, Cleary F, et al., editors. The Future Internet. Vol. 6656. Berlin, Heidelberg: Springer Berlin Heidelberg; 2011. pp. 431-446

[25] Chourabi H et al. Understanding Smart Cities: An Integrative Framework; 2012. pp. 2289-2297

[26] Zanella A, Bui N, Castellani A, Vangelista L, Zorzi M. Internet of things for smart cities. IEEE Internet of Things Journal. 2014;**1**(1):22-32

[27] Batty M et al. Smart cities of the future. The European Physical Journal Special Topics. 2012;**214**(1):481-518

[28] Hernández-Muñoz JM et al. Smart Cities at the Forefront of the Future Internet. In: Domingue J, Galis A, Gavras A, Zahariadis T, Lambert D, Cleary F, et al., editors. The Future Internet. Vol. 6656. Berlin, Heidelberg: Springer Berlin Heidelberg; 2011. pp. 447-462

[29] Pitarma R, Marques G, Ferreira BR. Monitoring indoor air quality for enhanced occupational health. Journal of Medical Systems. 2017;**41**(2)

[30] Molisch AF et al. IEEE 802.15. 4a channel model-final report. IEEE P802. 2004;**15**(04):0662

[31] Marques G, Pitarma R. Health informatics for indoor air quality monitoring. In: Information Systems and Technologies (CISTI), 2016 11th Iberian Conference on, 2016; pp. 1-6

[32] Marques G, Pitarma R. Smartphone application for enhanced indoor health environments. Journal of Information Systems Engineering & Management. 2016;**4**(1):9

[33] Marques G, Pitarma R. An indoor monitoring system for ambient assisted living based on internet of things architecture. International Journal of Environmental Research and Public Health. 2016;**13**(11):1152

[34] Marques G, Pitarma R. Monitoring Health Factors in Indoor Living Environments Using Internet of Things. In: Rocha Á, Correia AM, Adeli H, Reis LP, Costanzo S, editors. Recent Advances in Information Systems and Technologies. Vol. 570. Cham: Springer International Publishing; 2017. pp. 785-794

[35] Marques G, Pitarma R. IAQ evaluation using an IoT CO_2 monitoring system for enhanced living environments. In: Rocha Á, Adeli H, Reis LP, Costanzo S, editors. Trends and

Advances in Information Systems and Technologies. Vol. 746. Cham: Springer International Publishing; 2018. pp. 1169-1177

[36] Yu T-C et al. Wireless sensor networks for indoor air quality monitoring. Medical Engineering & Physics. 2013;**35**(2):231-235

[37] Myers SS et al. Increasing CO_2 threatens human nutrition. Nature. 2014;**510**(7503):139-142

[38] Neuburg M. iOS 7 Programming Fundamentals: Objective-c, xcode, and cocoa basics. United States: O'Reilly Media, Inc.; 2013

[39] Lacis AA, Schmidt GA, Rind D, Ruedy RA. Atmospheric CO_2: Principal control Knob governing earth's temperature. Science. 2010;**330**(6002): 356-359

[40] Awbi HB. Ventilation of Buildings. Taylor & Francis; 2003

[41] Marques G, Roque Ferreira C, Pitarma R. A system based on the internet of things for real-time particle monitoring in buildings. International Journal of Environmental Research and Public Health. 2018;**15**(4):821

[42] Smith A. Pew Research Center. Smartphone Ownership–2013 Update. Vol. 5; 2013

[43] van Deursen AJAM, Bolle CL, Hegner SM, Kommers PAM. Modeling habitual and addictive smartphone behavior. Computers in Human Behavior. 2015;**45**:411-420

[44] Hintze D, Findling RD, Scholz S, Mayrhofer R. Mobile Device Usage Characteristics: The Effect of Context and Form Factor on Locked and Unlocked Usage; 2014. pp. 105-114

[45] Rahmati A, Zhong L. Studying smartphone usage: Lessons from a four-month field study. IEEE Transactions on Mobile Computing. 2013;**12**(7): 1417-1427

Chapter 4

Best Illumination Scenes for Spaces Users

Naglaa Sami AbdelAziz Mahmoud

Abstract

Can we live in a dark environment? Light is the essential element, natural or artificial, traditional or sustainable, that helps us proceed in our life. Creating lighting scenes is one of the important roles of an interior designer, to create the interior environment for the users, whether in private or public spaces. Designing appropriate lighting to the function, the designer refers to the ideal set design using artificial elements in addition to the possible natural penetration to reach the complete lighting scene, which suits the type of interior function. The lighting design differs from interior type to others, and success of the lighting scene contributes to the success of the full experience of all the places we live. This chapter will explore the possible lighting design that affects positively on the life enhancement, as a physical and psychological tool, of most of the interior types.

Keywords: lighting design, interior design, lighting scenes for interiors, artificial lighting design, natural lighting design, lighting psychology design

1. Introduction

Illumination, or to be under the light, is a phenomenon that normally happens under the natural light. Every morning, with the sun rising, all its surroundings illuminate. For a long time, natural lighting was the essential tool to see what surrounds us, until the discovery of the fire, which remained the main light source until 1879 where the artificial light bulb started being used commonly.

A long time before civilization, the illumination process was only sustainable, since it used natural resources that occur without human intervention. The natural light does have its regulations as it appears in specific positions, which change over day times. To reach, a particular scene, using natural lighting, several solutions are possible, where some could be classified under sustainable solutions, and scientists along with designers have created new systems repeatedly. The tools that control, both natural and artificial lighting, are diverse but enable the interior lighting designer to create a particular scene that affects positively on spaces' users, to improve their quality of life [1].

Like many discoveries, scientific developments, and human creations, the specialization became an urge within any profession. Thus, the interior designer nowadays is much keen to comply with the users' needs, as many details are considered for his well-being, physical and mental health, and his entire safe life experience—as per the declaration of the International Federation of Interior Architects/Designers (IFI). The architect has his responsibility in the complexity of the buildings' layouts. However, the architect, while interfering in the interior space, as he used to do for many decades before 1880, creates such sterilized interior lighting that could fit in many interior types, no matter the user's natures, needs, nor wishes. The results are usually boring, with the absence of any identity of those inhabitants, and with no care of the user's moods and desires [2].

Lighting design is a specialization within interior design, which has effects over the overall human life experience. By changing its density, its colors, and its positions, the interior lighting designer could create unlimited scenes that fit with the users' requirements—the power of light to transform any space—as per the International Association of Lighting Designers (IALD). This power of alteration affects both the physical and the psychological aspects of all humankind. These effects occur unconsciously; once the person is in a specific environment, he starts to react based on its cognitive and emotional responses to this environment [3].

2. Types of lighting and spaces' functions

A long time before any civilization, humans used to live surrounded by nature, using the available resources given to them naturally. Each day, by sunrise, people indulge in their daily life, as a normal reaction to the natural phenomena. They spend their day working, and by the sunset, they start to feel unsecured as opposite to the day feelings. As time passes and by coincidence, they discover the fire that promoted their security as well as an extension to their days [4].

Many decades after this discovery, by 1880, and due to the development of sciences, knowledge, and life, the electricity and the electrical bulb occur in humanity. As all sciences keep developing, the discoveries in favor of the human health, safety, and comfort show up constantly and even became daily. These inventions are in favor of the human, but on the other hand, because the human creates them, they do have side effects. Only God's creations are safe, healthy, and comfortable for the human.

Each type of bulb is created for the people' benefits; after time and through their usage, science discovers their side effects and then tries to find the appropriate solutions to overcome these circumventions. For the past decade, some types of artificial lighting sources developed to overcome the previous version. The inventions reached the substitution of natural light by pure, sustainable lighting that could grow natural plants within it. Science did not discover the side effects of this great addition to humanity, from the scientists' point of view, but from the user ones.

The great polemic using either natural or artificial lighting is in its pick. The innovative sustainable systems that allow the consumptions of natural light open unlimited opportunities to the designers for additional sources of conceptions. Artificial "sustainable" lighting reaches a diversity that enriches the possibilities for the interior lighting designers in their duties for the favor of humanity. Also, this wide range of artificial light source options helps the designers to create the suitable visual environment tool to reach a convenient interior; this depends on the type of function in a specific case [5].

2.1 Artificial-natural lighting and interiors

In 1875, Henry Woodward copyrighted the first electric light bulb. While in 1876, Pavel Yablochkov invents the Yablochkov candle, the first practical carbon arc lamp, for public street lighting in Paris. Finally, in 1879 Thomas Edison and Joseph Wilson Swan patent the carbon-thread incandescent lamp. It lasted for 40 hours, while today the light-emitting diode (LED) lamps could last for up to 12 years. Between these origins and the latest LEDs, many families of lamps appear. For each type of bulb, after a certain time of usage, side effects start to appear, and therefore the scientists take responsibility to overcome their problems by healthy solutions as well as by developing the bulbs themselves. Each time, the argumentation reaches the point of what are the benefits of the artificial light and the natural light.

2.1.1 Artificial light sources

The first group, the one which starts with Edison's incandescent, were the tungsten bulb and the halogen bulb. Both produce light by heating the tungsten filament, using the electrical circuit. The light produced is warm and gives the same feeling of the sunlight but with less brightness than natural illumination. Heating the tungsten needs a lot of electricity, and by several usages, the filament's life ends, in addition to the extra load of heat provided. This type of source is great for the residential and especially in the reception areas as they provide the warm, welcoming feelings (**Figure 1**). They are also great in the high-level restaurants as they do give the appetizing feeling to the food (**Figure 2**) [6].

The second family is the gaseous group and consists of two subgroups: the low-intensity discharge (LID) and the high-intensity discharge (HID). The LID, famous enough with the popular light tube, the fluorescent, and then the more popular safe-energy lamps are in the everyday usages nowadays. They need special electrical systems yet less expensive and much longer in terms of lifetime. The quality of light produces pale and bluish yet efficient in terms of electricity. These tubes and bulbs are the best in commercial low-budget places. Ordinary offices, factories, schools, and universities are among the top interiors that use them despite the bad effects on the human skin if used away from some warm illuminations.

The second subgroup in the second family is the HID. Many lamps belong to this subgroup. The most effective for the interior functions are the metal halide (bulb shape) and the cold cathode (as the most flexible lamp, for the cove integration

Figure 1.
Reception. Source: Google Search—Images.

Figure 2.
Restaurant. Source: Google Search—Images.

Figure 3.
Shop. Source: Google Search—Images.

system to fit any shape, any color of light in a very limited space). Metal halide lamps are bright and are more efficient concerning the lifetime and the electricity consumptions. Elegant public spaces use them widely (**Figure 3**), the same as for the cold cathode or the light cove tube that fits in and affords adequate illuminations for the leveled public interior (**Figure 3**).

Finally, the electrical family of artificial light sources consists of the LED and the organic LED. In addition to their properties that exceed the electrical efficiency, their light is safer to the users regarding harmful rays and heat production, so the museum is the ideal interiors to rely on these illumination sources (**Figure 4**). The

Figure 4.
Museum. Source: Google Search—Images.

flexible OLED that came in shape, especially the OLED that came in shape of a thinner plate of light, is a real evolution of the artificial light sources. As an extended lifetime light source, they are the sustainable ones. Therefore their applications in private spaces are more than public spaces, yet depending on the level of awareness of the clients, they are the expensive type concerning their installation, but they are the most economic concerning their consumption of electricity and their lifetime.

2.1.2 Natural light source

As a natural phenomenon, the earth moves around the sun at very high speed, the fact that makes us feel stable, and we used to say "Sun Movement". The natural light starts with the sun rising of the sun, reflecting over the environment and affecting the universe. Natural light then splits to direct sunlight and simply daylight. The natural light benefits depend on the location toward its path through the day.

In general, the advantages of the natural illumination on a human, in addition to the energy saving, are the soft distribution of light that reveals the true colors and the enhancement of the visual acuity. The natural light ensures the security feelings in all human kinds. Moreover, the natural light helps to add exterior views to the interior that promote a direct link to the alive movement that stimulates the time feelings and prevents the seasonal affective disorder (SAD), which is the most mental effect of the absence of natural lighting. As a conclusion, the natural light enhances the life productivity [7].

On the other hand, the disadvantages of the natural light are associated with the penetration of excessive glare, the ultraviolet rays (skin health problems and color fading effects) and noise, the lack of privacy, the heat gain especially in summer, and maintaining the cleanliness level. In general, the solutions that face these problems are many and suit the different needs and situations [8].

Early morning, the sun rises from the east giving bright luminous light (spring and autumn, from east to north in summer, and from east to south in winter). At the midday, the sun is toward the south, affecting the interiors with different angles depending on the season and our position on the globe. The light is warm and vivid (around 45° in mid-seasons, less in summer about the 60°, and more in winter reaching the levels of the 30° to give more benefits to the users). Then in the afternoon timing, the sunlight faces the west with radiant reddish light (spring and autumn, from west to north in summer, and from west to south in winter).

Therefore, the sun movement affects the interiors depending on a different time with variable color and intensity. As an example of interior functions and its relations to the natural illumination, the selection of the function position is crucial to enhance the human life. As an example, the residential case will clarify the best position and location of the interior functions as the following:

- The bedrooms should face the east so in the early mornings the sunlight creates a vivid push to the users to start their day, fresh and full of energy, (**Figure 5**). Also, the health benefits that result from the direct sunrays on the sleeping surface are enhancing the vitamin D absorptions (the best time to expose the human bodies) and confronting the bedsheets' microbes.

- Kitchen (**Figure 6**) and hobby areas facing the south direction will improve the activities done, giving the linkage feeling to the exteriors. Nature promotes the housewives' lives in a great way!

- The afternoon soft, warm, and cozy illuminations indorse the living areas (**Figure 7**) where the family gathers after the lively activities. Therefore, these areas facing the east are the best positioning.

- In the special area where artists practice, the best location would be facing the north. The north never has direct sunlight, and therefore the level of shades is soft, creating a uniform level of illumination to the artistic productivity (**Figure 8**).

Figure 5.
Bedroom. Source: Google Search—Images.

Figure 6.
Kitchen. Source: Google Search—Images.

Figure 7.
Living room. Source: Google Search—Images.

These natural lighting scene examples are the ideal configurations of the sunlight benefits on human activities and therefore its productivities taking into considerations its physical and mental benefits [9].

2.2 Sustainable lighting and interiors

As previously discussed, vis-à-vis the responsibility of the interior design over the human's well-being, sustaining the natural resources becomes a master duty. Our planet grows in population, and the diminution of natural resources increases. Designers and industrials play an important role, today, to create solutions that

Figure 8.
Artist studio. Source: Google Search—Images.

counteract this global and critical problem. Furthermore, users need to have a wider perspective and an appropriate awareness, to use these innovative solutions in favor of pertaining the remaining natural power.

The interior lighting designer needs to have the full scope of possibilities, to start integrating them into the interior design. These sustainable lighting tools represented devices and creative design systems.

2.2.1 Light sources

LED lamps are the top sources of artificial lighting. They are available in all range of white and colored light tones. As they depend on electricity composition, they produce the most neutral balanced white light that helps in the lighting sets in favor of the users.

2.2.2 Light systems

The main strategy of nowadays interior lighting designers is to exploit to the maximum the available products for a sustainable methodology. The innovative systems, created by the scientists, help in creating a convenient interior environment for the benefits of the users, in conditions of their positive applications. The results are different lighting scenes, natural and artificial, under the canopy of sustainability. These systems are:

- Light pipes: (**Figure 9**) consists of the main high reflective surface at the top of the building, connected to a pipe tube, ending on the point of needed light. The composition of the inner surfaces of the full connection should be from highly reflective surfaces. A tiny solar panel could be joint, facing the solar energy, and function as an extra natural-artificial light source at nighttime. The resulting lighting scenes are not diverse.

- Solar tracking system: (**Figure 10**) comprises a set of moveable mirroring materials that track the sun movement and then reflect it back to the interiors. This system increases the natural lighting penetration in the interiors, promoting variable lighting scenes for the benefits of the occupants.

- Lighting shelf: (**Figure 11**) it is a very simple tool to add more reflective natural illumination and prevent the excessive glare from the direct sunray. It consists of glossy white shelf added to an approximate level from the ceiling while facing the sunray. It is near to the dynamic louvers, which provides the reflected natural lighting when needed, and close it totally if not needed. The difference between the two systems is that the fixing system of the second one is from the outside not from the inside as the first one. The dynamic louvers create variable illumination scenes based on the users' needs and desires for the favor of the function.

- Sun scoop system: (**Figure 12**) it is so similar to the solar tracking system, where the only difference is that they fixed the mirroring materials. Both of them add natural lighting to the nearest level of the roof [10].

Figure 9.
Solar light pipes system. Source: Google Search—Images.

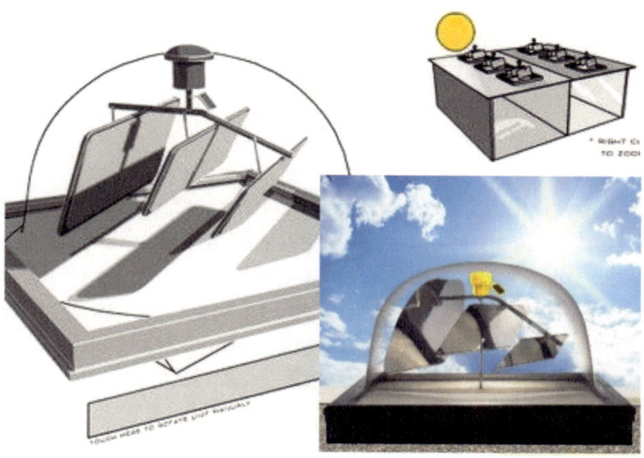

Figure 10.
Solar tracking system. Source: https://www.treehugger.com/interior-design/solar-tracking-skylights.html.

Figure 11.
Light shelf and dynamic façades systems. Source: Google Search—Images.

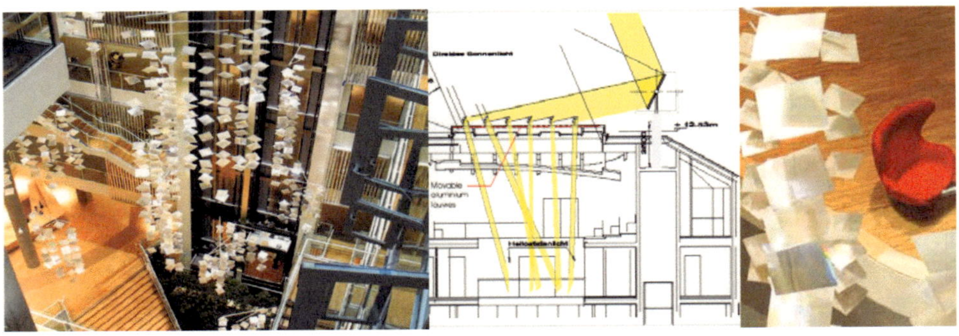

Figure 12.
Sun scoop system. Source: Google Search—Images.

3. The interior lighting designer role

The interior lighting designer, today, has an ethical obligation to protect the health, safety, and the welfare of the interior inhabitants. The expansion of knowledge required the respect of the specializations studied at the undergraduate level, to prepare the future professionals with the knowledge and applications needed to face the real developed life. Interior Design became a multifaceted profession, where many specializations are available under its umbrella and need a special focus to study its aspects for producing professional designers. With the growth of specialization in all kinds of knowledge, one of the specialized fields became more complex and needed particular focus, the lighting design [11].

3.1 Interior designer vs. architect

Respecting the specialization in all fields is important. The more science develops, the more details appear, as each profession should respect the other. The architect focuses on the standards, as they are responsible for much detail concerning the complexity of the buildings. While the interior designer with strong lighting design background creates the lighting that illuminates the users based on the physical needs and requirements as well as the mental and psychological necessities and effects.

A simple comparison as in **Figure 13** shows two different lighting scenes for the same interiors, between lighting applications by architects and lighting design by interior lighting designer. The warmer the residential space, the better it is for the

Figure 13.
Two different locations in residential interiors (architects vs. interiors designs). Source: Google Search—Images.

users. The couples [1] are for the same kitchen where applying the users' needs and taking in considerations the psychological effects of all the interior elements, shows the lighting scene enhancement happened to the same interior, on the right side from both interiors. The couples [2] are the same small lobby where the two lighting scenes show the difference of applications.

In general, when the architect adds the standards, the interior appears under sterile conditions. The psychological studies for the favor of the specific users compliment the message that these users want from their spaces. The studies of the entire design program promote the creation of interior and especially the lighting setting that enhances the lives of these users, as the design decisions are part of their inputs.

4. Philosophy of lighting scenes

Lighting design creates the physical needs as visibility for different tasks. It ensures the psychological needs of the visual comfort, mood, atmosphere, health, and safety. The philosophy behind designing the lighting scenes, or the lighting scenography, is to stimulate the environmental conditions of the light but in a manner to suit the specific needs in time and position. The main philosophy of the natural light is that it is variable. Every 3 minutes, the natural light changes in intensity, color, and direction. The natural light occurs in irregular scenes over the day, which is the essence and the philosophy behind that any designer seeks to realize in its lighting design.

The human is constantly under changeable lighting scene, in nature, that becomes the goal of any designer to realize. If the artificial light remains fixed, the full environment leads to boredom and sterility. To reach such sets, industrials with the consultancy of designers created the tools and systems to change in controllable ways the sequential changes needed in the lighting in safe, economical, and user-friendly tools [12].

4.1 Lighting scene tools

In the scenography, or the lighting scenes, several tools were available to ensure the purposed functions. Among these tools, the primary one is the switches. Switch devices vary from basic on/off to complicated devices that suit the innovated technologies of lamps and integrated lighting fixtures. The lighting scene design or the lighting scenography refers in some settings to the usage of colored light. Additionally, the lighting scenography, when involving a number of lighting fixtures, should recall for a programming system to ensure its functionality.

4.1.1 Switches

The primary tool to control the lighting design is the switches. Simply by switching on/off, we create two different scenes. These switches, if appropriate to the type of lamps or luminaires, could involve the dimmer properties. Once the dimmer is used, many scenes are created in the interior. The availability, only, of switches/dimmers, could produce unlimited variations of lighting scenography in a condition that the primary light distribution meets the users' needs and mood.

4.1.2 Color and scenography

Colored light is best produced by red, green, and blue (RGB) color mixing luminaires or by using color filters. The use of different color temperatures and colored light augments the spatial emphasis on specific objects. The uses of saturated colors help in the creation of intense environmental lighting effects.

4.1.3 Programming the scenography

A lighting program is a tool that controls a larger number of groups of luminaires. It allows light scenes to be recalled at the touch of a button. The program permits to take place over a given period. It allows controlling the light by daylight (using sensors) [13].

4.2 Analysis of lighting scenes and function

As any space consists of a three-dimensional layout, lighting scenes approaches should consider more visual textures by using layers of light. The more depth created, the more varieties in interior perceptions allow the vibrant interior experiences.

We all live in a place over the day. These spaces when fixed became a routine that leads to boring feelings and after time reduce the human energy. To keep the life challenging, an analytical approach to the main type of interiors will share in creating such a challenging life experience. Through the types of interiors, we all use houses, exhibitions, offices, and school or universities in daily consistency [14].

4.2.1 Residential interior scenography

Throughout the world, in capitals, cities, or suburbs, residential interiors encompass a large variety of structure compositions. Residences are available in an extremely broad range of sizes and in an amazing array of configurations [5].

The complexity increases when adding a variety of the interior components such as furniture, textures and surfaces, color schemes, equipment, and accessories, which are unique in each residence. In **Figure 14** is a single living area analysis through some example of lighting scenography:

- First (1) scene: The full space is completely unlit. An eye-catching feature is the fireplace.

- Second (2) scene: The bright luminous ceiling produces a scene analogous to an overcast sky bringing diffuse light into the room. Additional wall washing allows the textured wall to be recognized.

- Third (3) scene: More dark ambient, where only the fireplace plays as a focal point and the two narrow beam pendants with extra focus on the greenery.

4.2.2 Office interior scenography

Lighting design in the office follows the functional approach in workstation areas but depends on the message and the type of business that need to appear to the clients. Philosophy of design plays major role in creating the multiple scenes needed in such spaces, especially in the reception area, exhibition area, and in the meeting rooms, all along the productivity quality of the workable area. In **Figure 15**, an example of a reception—exhibition area of an office shows the possible scenes which allow the company the professional presentation:

- First (1) scene: Accent light in combination with wall washers produces a balanced and differentiated lighting solution. The lighting supports the different functional zones of reception, small and large displays.
- Second (2) scene: Effectively stage the exhibits and create rich contrasts by using recessed spotlights.
- Third (3) scene: The uniform vertical illumination creates depth and provides a calm background. The wall washer is arranged on the ceiling and underneath the gallery.

4.2.3 Exhibition and scenography

Exhibition design, (**Figure 16**) especially the temporary setting where the artwork in the hall emphasizes in subtle ways the uses of only slight contrast with the surroundings or set off in rich contrast using sharp-edged projections, will create the end appearance. The following scenes could help in creating some attraction to the medium to long exhibition timing:

- First (1) scene: Uniform lighting sets to the overall components of the space, including the space layout.
- Second (2) scene: Only uniform vertical illuminance over the main orange artwork.
- Third (3) scene: The directed light highlights the sculptured element.
- Forth (4) scene: The artworks on the rear wall illuminated by nonuniform lighting.

In the case of color light interfaces, the ambiance of scene design changes in its setting as (**Figure 17**):

Figure 14.
Different lighting scenes (residential) using simple switch/dimmer. Source: ERCO guide and catalog. https://www.erco.com/guide/indoor-lighting/planning-examples-5867/en_us/.

Figure 15.
Different lighting scenes (offices). Source: ERCO guide and catalog. https://www.erco.com/guide/indoor-lighting/planning-examples-5867/en_us/.

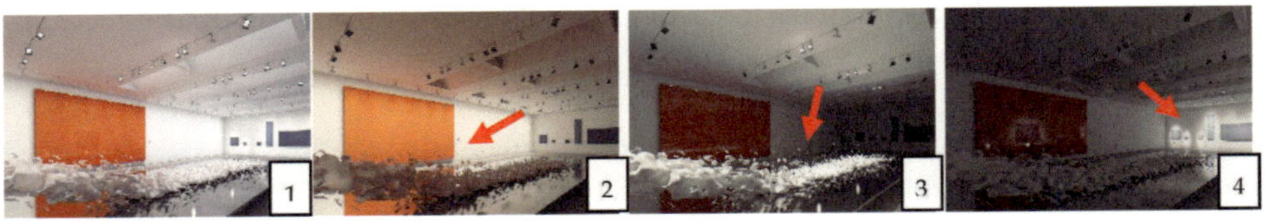

Figure 16.
Different lighting scenes (exhibitions). Source: ERCO guide and catalog. https://www.erco.com/guide/indoor-lighting/planning-examples-5867/en_us/.

- First (1) scene: When the lighting design relies exclusively on white light, the brightness levels produce the differentiating contrast. The focuses on the three-dimensional object enhance the display.

- Second (2) scene: Using a cold light color intensifies perspective and creates an open feeling of space for the objects accentuated with warm white light.

- Third (3) scene: Using colored light can alter the appearance of white objects and transform the neutral-colored room concept into a world of intensive color.

4.2.4 Classrooms and lighting scenes

In the classroom, where the educational experience occurs, the lighting should follow the efficiency philosophy and promote the educational strategies to support the learning philosophies. Quality of light, curiosity, and an attractive environment are key design tips for these specific interiors. **Figure 18** will give an idea of lighting scene in the classroom:

- First group of scenes: The natural light and the artificial light play together for creating different appearances related to the functions needed.

- Second group of scenes: The artificial light with different intensities, focuses, and directions again to suit different activities in the same classroom.

4.3 Additional lighting scene as life enhancement

The environmental impacts that have the light on the users and their surroundings exceed the control of the light scenes or the creation of these scenographical lighting sets. In addition to the importance of creating multiple lighting scenes that break the monotony of any spaces, it is part of the natural phenomenon that any lighting designer seeks in their creations. It is clear that each interior could hold some light scenes that will enrich its functions and will excite its users in favor of their duties.

Figure 17.
Different lighting colored scenography in a temporary exhibition. Source: ERCO guide and catalog. https://www.erco.com/guide/indoor-lighting/planning-examples-5867/en_us/.

Figure 18.
Different lighting scenes for a classroom. Source: Google Search—Images.

Additional regulations could help the lighting designers, when the specific environment is in the request, no matter the type of interior. Such additional elements could reach with any interior to a level of high perfection, yet nothing could be complete.

These specific ambiances could appear as an interior as pleasant, public, spacious, relaxing, or interior that promotes privacy. The following guidelines help to create these special sets of lighting design [15].

4.3.1 Pleasant interiors

Places like homes and hospitality, where pleasant illumination can endorse the psychological effects of their users, especially in hotels and restaurants, so clients can remember their pleasant experience that will promote their turn back to the place more than one time. To reach these feelings, use wall wash lighting (**Figure 19**), instead of directly down from the ceiling. In addition to the uses of the accent, nonuniform distributions of brightness in the space.

4.3.2 Public interiors

All public spaces should respect the social and public proxemics. For some interiors, the physical clearance is hard to reach due to some considerations, like the spaces' expenses. In such examples and others, designers seek to add the public feelings to the users. The public interiors should respect the individualism as strangers are surrounding each other, so the secure feeling is necessary.

If a person feels unsecured, he will be under one of two options: fight or flight. Due to the limited personal space, once a stranger intrigue in his bubble and the escape way is not clear, they start fighting each other. In the case of a clear path to the outer space, the person who feels in the trap will fly away from this specific space. To confront the limitations in public, lighting designers have to use high levels of illumination with more distribution of light from overhead lighting sources to provoke these senses (**Figure 20**). Strangers need to see each other's faces to augment their feelings of security.

4.3.3 Spacious interiors

Open spaces give human the senses of security, especially in the daytime under the direct sunlight. As he used to be, the human being needs to feel in a spacious environment. Again, the personal proxemics will take part in any interior design. The interior designer is responsible for people's comfort as well as to secure their physical needs.

For other situations, when the public function, like lobbies, is limited in dimensions yet receive some visitors, the lighting design could help to create a feeling of spaciousness by providing high levels of illumination and uniform distribution lighting on all surfaces (**Figure 21**). Horizontal luminance modifies the impressions of spaciousness, especially when focus on low ceiling heights.

4.3.4 Relax interiors

In our heavy lives full of duties, each needs to take a breath, particularly in his/her home. Many consider that just by lowering the light levels, the relax feeling will occur. Well in contrary, a specific lighting scene is necessary for such condition. Further to the pleasant interior conditions, a nonuniform distribution is essential. Moreover, the wall washer lighting or lower light levels distributed all over the place will create this relax feelings (**Figure 22**). Finally, the warm light level that stimulates the sunlight increases the relax feelings as it reminds the user of being under the sun on vacations [16].

4.3.5 Interiors that promote privacy

Private space suggests separating people, so each feel at ease. When a human is in public but need to practice some private experiences, like being in a restaurant

Figure 19.
Lighting scene for pleasant interior. Source: Google Search—Images.

Figure 20.
Lighting scene for public spaces. Source: Google Search—Images.

Figure 21.
Spacious interiors. Source: Google Search—Images.

Figure 22.
Relax interiors. Source: Google Search—Images.

Figure 23.
Lighting for privacy. Source: Google Search—Images.

surrounded by strangers and needing to eat in peace, the island of the light scene is the best tool to create the privacy within public (**Figure 23**).

Shadow and silhouette reinforce feelings of privacy. Moreover, the quantity and nonuniform lighting prevent the details of faces, from a distance, while the spot of light centered at each distribution setting will reinforce the intimacy feeling between the same spot [17].

5. Conclusion

- Originally, natural lighting is the sustainable lighting ever created. It was the main source where people used to run through their life.
- As natural light is variable regarding quantity, color, and intensity throughout the day and the year, the wise selection of functions in interiors about these restrictions should be wise to enhance human life.
- The more the lighting designer uses the sustainable systems to introduce the natural light into the interiors, the more audience will be familiar with applying them. The available variety could fit any specific interior and architectural layout.
- From the wide range of artificial light sources, the sustainable LED and OLED should remain the first choice when designing. It provides the best intensity, the appropriate color temperature, and the distribution for any specific interiors.
- The respect of specialization in all aspects of knowledge promotes its development in humanity favor, so the respect of the interior lighting designer is the key to success to produce the best illumination for the people. Architects do have their responsibility, so it is time to give space to the designer to carry the human life.
- Best illuminations scenes for places' users start by the tools to control them, the exploration of the importance to have such scenography, and the additional tips to produce them, ideally. Each type of interior has a specific guideline in the courtesy of the function, in addition to the users and space layout and design.
- Lighting levels and lighting color affect our behavior and our perception of interiors.
- The impression of space results from the relationship between lighted surfaces and surroundings (focus and background), to provide public lighting scene, or pleasant, or private, or even spacious scenography.
- Standard lighting design lead to sterile environments. If all objects and surfaces in an interior receive the same importance of illumination, over time, the occupant will feel depressed. Assuring the brightness contrast will lead to an inviting and inspiring interior.
- If all objects and surfaces in an interior receive the same importance of illumination, over time, the occupant will feel depressed, so the changeable variations in the lighting scene will help to fight this negative feeling.
- Controlling the brightness contrast results in an inviting and inspiring interior.

Nomenclature

ftcd	foot-candle is the level of light. Outdoor varies from 10,000 to 10 ftcd (sunlight to very dark days). Indoor varies from 20 to 10,000 ftcd (public areas with low focus to health-care operation rooms)
lm	lumens, measuring unit for the quantity of light output from the light source or the lamp
w	watts, measure the quantity of power (electricity) needed by a lamp to provide light
K	kelvin, the measuring unit of the color temperature of light
CRI	color rendering index is the unit that shows the level of appearance of the object under a specific lamp
lx	is the luminous intensity based on the quantity of illumination
LID	low-intensity discharge, group of lamp including "fluorescent" and "CFL—compact fluorescent light"
CFL	compact fluorescent light
HID	high-intensity discharge, group of lamp including "metal halide" and "cold cathode"
LED	light-emitting diodes, lamp type based on electricity technology
OLED	organic LED, the last invention of lamps
RGB	red, green, and blue. They are the principle color of light

Author details

Naglaa Sami AbdelAziz Mahmoud
Interior Design Department, Ajman University, Ajman, UAE

*Address all correspondence to: n.abdelaziz@ajman.ac.ae

References

[1] Steffy G. Architectural Lighting Design. 3rd ed. New York: Wiley; 2008. ISBN: 978-0470112496

[2] Advocacy Platform: https://ifiworld.org/ [Accessed: March 10, 2018]

[3] The Power of Light: https://www.iald.org/ [Accessed: May 9, 2018]

[4] Brandi U. Lighting Design; Principles, Implementation, Case Studies. Munich, Germany: Birkhäuser Edition; 2006. ISBN: 978-3764374938

[5] Bacon L. Interior Designer's Attitudes toward Sustainable Interior Design Practices and Barriers Encountered when Using Sustainable Interior Design Practices [thesis]. Lincoln: Architecture Program: University of Nebraska; 2011

[6] Descottes H, Ramos C. Architectural Lighting: Designing with Light and Space. New York: Princeton Architectural Press; 2011. ISBN: 978-1568989389

[7] Environmental Health Perspectives: https://www.ncbi.nlm.nih.gov/pmc/articles/PMC2290997/ [Accessed: May 10, 2018]

[8] Fowles D. Interior design education in the year 2000: A challenge to change. Journal of Interior Design. 2008;**17**(2):17-24

[9] Sami Abdelaziz N. Specialization in teaching the lighting design courses between the interior design program and the architectural engineering program. In: 6th International Technology, Education and Development Conference; 5-7 March 2012; Valencia, Spain; 2012. pp. 2875-2883. ISBN: 978-84-615-5563-5

[10] Bergman D. Sustainable Design: A Critical Guide. New York: Princeton Architectural Press; 2010. ISBN: 978-1568989419

[11] Brooker G. What is Interior Design? Rockport Publishers; 2010. ISBN: 978-2888930174

[12] Winchip SM. Fundamentals of Lighting. 3rd ed. New York: Fairchild Books–Bloomsbury; 2017. ISBN: 978-1-5013-1766-8

[13] Martin L. The Lighting Bible–Ideas for Every Room in Your Home. Page One–Quarto Publishing; 2011. ISBN: 978-981-245-743-1

[14] Discovering Light–Design Examples: http://www.erco.com/guide/indoor-lighting/planning-examples-5867/en_us/ [Accessed: May 20, 2018]

[15] Sami Abdelaziz N. Linking and applying 'psychology of design and interior design' courses to enhance spaces functionality and aesthetics. In: 5th International Conference on Education and New Learning Technologies; 1-3 July 2013; Barcelona, Spain; 2013. pp. 1162-1170. ISBN: 978-84-616-3822-2

[16] Colombo B, Laddagaa S, Antonietta A. Psychology, and design. The influence of the environment's representation of emotion and cognition. In: 6th International Conference on Applied Human Factors and Ergonomics (AHFE); 2015. DOI: 10.1016/j.promfg.2015.07.370

[17] Lighting Nomenclature and Definitions: https://www.nap.edu/read/24619/chapter/11 [Accessed:August 17, 2018]

Chapter 5

Spatial Distribution of the Nature of Indoor Environmental Quality in Hospital Ward Buildings in Nigeria

Pontip Stephen Nimlyat, John James Anumah, Michael Chijioke Odoala and Gideon Koyan Benjamin

Abstract

This study seeks to ascertain the spatial distribution of IEQ in hospital wards based on the physical measurement of the hospital ward units with different architectural features. Field survey was undertaken in the medical and surgical wards units of two case study Hospitals both located in Jos, Nigeria. IEQ parameter variables were monitored and recorded, and compared against recommended international standards for hospital facilities. Results show that the measurements of the IEQ parameters conditions in the selected hospital ward buildings, differ substantially depending on the ward design configuration and orientation, and also the outdoor weather condition. The indoor environment in the hospital wards had different thermal conditions because of variations in orientations, window sizes and air inlet/outlet. Building orientation, also affected the indoor daylight quality in each of the hospital ward buildings within the period of measurements. The Teaching Hospital wards whose orientation (NW-SE) allows the fenestration façade to fall within the sun path, maximised it for daylighting within the wards. It is therefore recommended that, the design of hospital wards for improved IEQ conditions should be such that proper attention is given to the orientation, floor plan configuration and window design for natural ventilation and lighting.

Keywords: design configuration, orientation, spatial distribution, indoor environmental quality, hospital wards

1. Introduction

Indoor environmental quality (IEQ) has been defined as, the determination of the significant factors that have direct effect on a building occupant comfort and wellbeing [1]. IEQ is also seen as components responsible for an environment that appears to be psychosocially healthy for its inhabitants [2]. Buildings are designed and constructed generally for human habitation, as a result, the requirements for their usage is needed to be fulfilled as a precondition for their well-being [3]. Different studies on IEQ in buildings have shown that, achieving comfort in an indoor environment should be based on the assessment of thermal, acoustic, lighting and the quality of the indoor air as parameters of importance [3–7].

Indoor environmental quality assessment in buildings have taken different dimensions based on building type and environmental setting. For example, it has been asserted that the assessments of the overall indoor environmental quality of office buildings do not follow any standard protocol [8]. This is also typical of hospital facilities where little has been done in assessing their indoor environmental quality, which has great impact on occupants' health, productivity as well as building energy demand [6, 9]. There is a need therefore, to have a standard measure for the overall indoor environmental quality of hospital buildings in order to prevent the negative impact of the environment on building occupants. The design of the hospital building environment should therefore be such that it has positive influence on occupants' health, comfort, and productivity.

The central theme surrounding any particular hospital is 'patient care'. Therefore, the design of a hospital building should be such that the patient as the main occupant experience comfort and protection from environmental elements. The hospital is seen as a therapeutic environment for caring for the sick and other related activities such as learning and research [10]. Neglecting the quality of a hospital environment will amount to issues that contradict the essence of a hospital as a healing environment.

The hospital which is generally seen as an environment for healing could possibly be harmful to both people and the environment [11]. Besides, occupants of hospital environment could contract some healthcare acquired infections that might even result into death. The adoption of sustainable design in healthcare facilities have been focusing on creating a facility that is supportive of an improved patient care, staff wellbeing and productivity, and environmental friendliness. Patient care, staff wellbeing and productivity, and environmental friendliness is referred to as the "Healthcare's Triumvirate [12] with respect to sustainable design in healthcare services. Different studies [11, 13, 14] have shown that promoting green hospital buildings could result into solving a wide range of sustainability issues which are challenging to the environment. Green building rating systems such as LEED, BREEAM and Australian Green Star (GBCA) have made some contributions towards promoting sustainable healthcare facilities planning, design, and construction [15].

Cross ventilation and building energy performance are among the main consideration in the design of hospital facilities. In as much as ventilation strategy is influenced by environmental settings, its long run cost implication is enormous. However, the running cost of providing proper ventilation using artificial system should not be deterrent in ventilating special hospital units such as Intensive Care Unit (ICU), Special Child Birth Unit (SCBU) and operating theatre among others. Air conditioning systems in a building allows for flexibility in form and space management [16, 17], while natural ventilation on the other hand requires use of courtyards for it to be achieved which takes much space. The provision of natural ventilation in hospital buildings would allow patients, and staff contact with nature [18], however, it has been attributed with infection spread.

Sustainable development and design processes in healthcare facilities is being driven by suppleness, cost minimization, innovation and healing environment [19]. Hospital design requires a careful consideration of the individual spaces to be provided and the incorporation of the requirement for optimum indoor environment, which is more challenging when compared to other building types. This study therefore seeks to ascertain the spatial distribution of IEQ in hospital wards based on the physical measurement of the hospital ward units with different architectural features (building orientation, design configuration and windows placement).

2. Methodology

The environmental conditions of different hospital settings might not be the same since they would have differences in certain architectural features, different management system and source of finance, and also, differences in the services they provide. Therefore, the need to study the indoor environment of different categories or settings of hospital facilities. To achieve the goals of this research, the rationality behind the concept and context in which the methods are employed needed to be explained, which will facilitate the evaluation of the research outcome [20].

In Nigeria, there are three main categories of public healthcare settings namely: the primary, secondary and tertiary healthcare facility. For this study however, only two of the hospital categories were chosen (Tertiary—Teaching Hospital and Secondary—Specialist Hospital). The criteria used in the selection of the case study hospital wards is based on the differences in their orientation, spatial and physical conditions, and organisational settings. Field survey was only done in the medical and surgical wards units of the selected hospitals. Plateau Specialist Hospital and the Jos University Teaching Hospital both located in Jos, Nigeria were selected as case study. Both the Plateau Specialist Hospital and Jos University Teaching Hospital are government owned hospitals that were selected because of the ease of access and ease in eliciting the required information. The selection of these two different hospitals is in understanding the variations that might likely be found in the nature of the IEQ performance due to some of their architectural features (orientation and design configuration) and organisational differences. For simplification, the case study hospital wards are designated by their orientation as Teaching Hospital (NW-SE) and Specialist Hospital (NE-SW) orientations.

2.1 The study area

This study is carried out in Jos, Plateau State, which is the twelfth largest state in Nigeria. The State is located on latitude 9° 10' N and longitude 9° 45′ E in central Nigeria. It has a total land area of 30,913 square kilometres and an estimated population of over 3 million people based on the 2006 population census, and having a population density of 100 per square of a kilometre [21]. Plateau State derived her name from the fascinating table top rock formation known as Plateau whose altitude ranged from 1200 to 1829 m above sea level. The climatic condition of the State could be referred to as temperate if compared with the other parts of Nigeria, even though it is situated within the tropical region of the world. The state records an annual mean temperature of between 18 and 22°C, with a mean annual rainfall of 1317.5 mm in the lower part of the Plateau and 1460 mm on the Plateau top.

The two case study hospitals are located in Jos the Plateau State capital city. Jos is located on latitude 9° 55' N and longitude 8° 53′ E and at an altitude of about 1200 m above sea level [22]. Jos has an average monthly temperature that ranged from 20.3 to 24.7°C, with an average annual rainfall of about 1300 mm as shown in **Figure 1**. Jos experienced an average monthly relative humidity of 53.4%. This weather obtained from the web page of Nigerian Meteorological centre does not reflect the current weather situation in Jos, Nigeria as the outdoor field measurement results shown in **Table 1** is significantly at par with the sourced data.

The selection of the case study hospitals was based on their ward orientation and configurations. The ward buildings in Plateau Specialist Hospital faced the Northeast-Southwest orientation while those of Jos University hospital faced the Northwest-Southeast orientation.

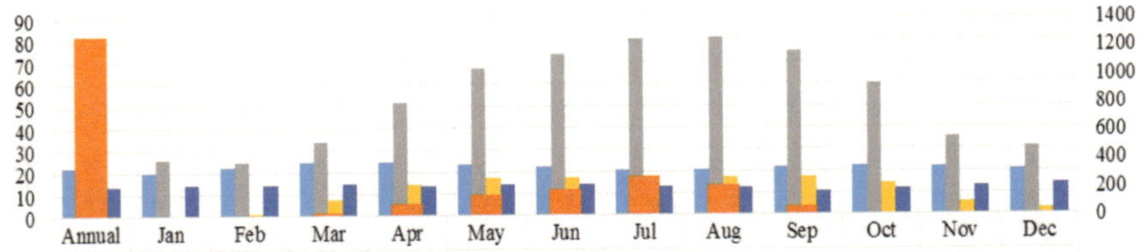

	Annual	Jan	Feb	Mar	Apr	May	Jun	Jul	Aug	Sep	Oct	Nov	Dec
Temperature (oC)	22.1	20.3	22.3	24.5	24.7	23.4	22.1	20.8	20.7	21.6	22.3	21.6	20.4
Relative Humidity (%)	53.4	25.8	24.5	33.9	51.8	67.5	73.6	80.5	81.5	74.9	60.1	35.7	31.2
Dew Point (oC)	12.2	0.1	1	7.6	14	17	17	17	17	17	14	5.8	2.8
Wind Speed (Km/h)	13.4	14.4	14.4	14.8	13.7	14	14	13.3	12.6	10.8	11.9	13.3	14
Rainfall (mm)	1285.9	1.1	4.8	23.8	83.2	153.9	185.1	274.9	214.9	63.8	3	1.4	0

Figure 1.
Summary of monthly weather averages in Jos. Source: NiMet [22].

	10 am	11 am	12 pm	1 pm	2 pm	3 pm	Average
Temperature (°C)	27.4	28.6	29.5	30.0	30.7	30.9	29.5
Wind (m/s)	2.5	2.3	2.0	1.4	1.4	1.4	1.9
Humidity	71%	61%	47%	44%	43%	41%	51%

Source: Field data.

Table 1.
Summary of daily hourly averages of outdoor weather in Jos (between 10 am and 3 pm, for 3 months period).

2.2 Plateau Specialist Hospital, Jos (NE-SW)

2.2.1 Hospital setting

The Specialist Hospital which represents a secondary healthcare facility is named 'Plateau Specialist Hospital' and is located in Jos, the State capital city (**Figures 2** and **5**). Plateau Specialist Hospital provides general and specialised medical services, and is also an accredited healthcare institution for residency in family medicine and internship training. The hospital has a bed capacity of 176 (124 adult and 52 children) and staff strength of 633 personnel, with an average daily patient flow or visits of about 176 per day [23]. The ward buildings selected in this hospital have a total bed space capacity of 64 (32 bed spaces each). Each of the ward buildings are partitioned into 6 (8) single rooms each, which can accommodate a maximum of 2 (2) inpatients.

2.2.2 Architectural features

The Specialist Hospital is located on latitude 9° 53′42.9″N and longitude 8° 53′02.2″E, and within a densely populated are of the city-centre. The hospital wards building layout and orientation as shown in **Figure 2** faces the Northeast-Southwest direction. The space organisation of the ward units is the corridor or continental form based on a description of hospital space types by James and Iatten-Brown [24]. The plan configuration of the hospital wards have 8 (16) units rooms of solid partitioned internal walls accommodating two bed-spaces each. The partitioned ward room units are access through a corridor (**Figure 3**) that separated them along two axis. Each of the ward room units is installed with two 1200 mm × 1200 mm Louvres glass windows on the same wall façade. The window to wall ratio (WWR) on the fenestration façade is 15%, which is less than the optimum recommended by Zain-Ahmed et al. [25]. The windows have curtains which were installed for shading.

Figure 2.
Site location of Specialist Hospital, Jos, Plateau State, Nigeria (Google Earth).

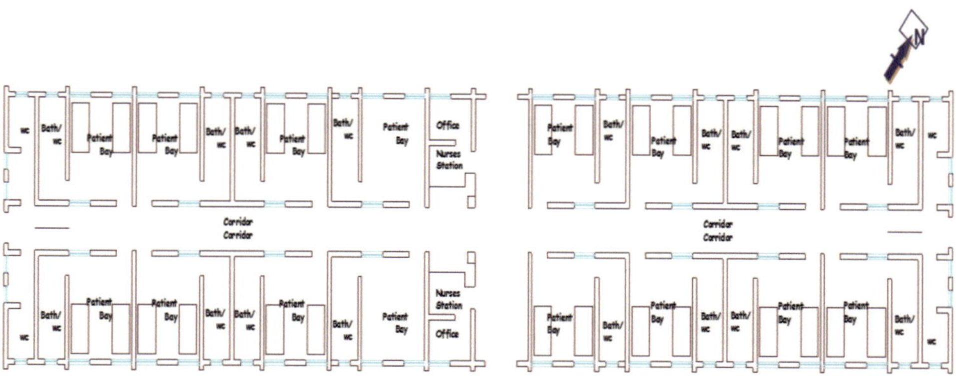

Figure 3.
Typical floor plan of Specialist Hospital ward.

The ward buildings are naturally ventilated with supplemental ceiling fans as artificial mechanical source. The mechanical source of ventilation (ceiling fans) powered by electricity are mostly not in use due to lack of power supply. A walkthrough observation of the hospital buildings revealed a lack of proper ventilation and lighting, and also, drain pipes leakages and fittings breakdown was observed. A typical floor plan and pictorial views of the Specialist Hospital is shown in **Figures 3–5**.

2.3 Jos University Teaching Hospital (NW-SE)

2.3.1 Hospital setting

Jos University Teaching Hospital was established in 1981 by an act of Nigeria Parliament, and was operating at a temporary site until 2007. Work on the permanent site was completed an inaugurated in May 2007 after operating at the temporary site for over 20 years. The permanent facility constructed for Jos University Teaching Hospital has a 620 bed-space capacity that provides both inpatient and outpatient services, as well as medical personnel training and research. The focus of this hospital is in providing tertiary health services training, using modern technology and research in a conducive environment for both patients and staff. The hospital facility is a 2 (2) storey building complex that houses all

Figure 4.
Specialist Hospital ward interior—(a) corridor, (b) typical ward room.

Figure 5.
Exterior views of Specialist Hospital ward—(a) northeast view, (b) southeast view.

departments, offices, research laboratories, and instructional classrooms. The selected ward buildings in this hospital are located on the first and second floor of the complex (**Figures 6–9**).

2.3.2 Architectural features

The Teaching Hospital is located on latitude 9° 54′27.5″N and longitude 8° 57′37.5″E, in a sparsely populated area at the outskirt 14.3 Km away from the city-centre. The hospital complex was designed by Interstate Architects Limited in the late 1970s with an initial size of 320 beds. The hospital layout was designed to expand to a 1000 beds Teaching Hospital at the appropriate time in future. The first phase of construction work of the complex was completed in 2006. The ward buildings are both naturally and mechanically ventilated with ceiling fan, and also, with split-level air-conditioning system which is often not in use. This facility has three sources of power supply which enable uninterrupted electricity

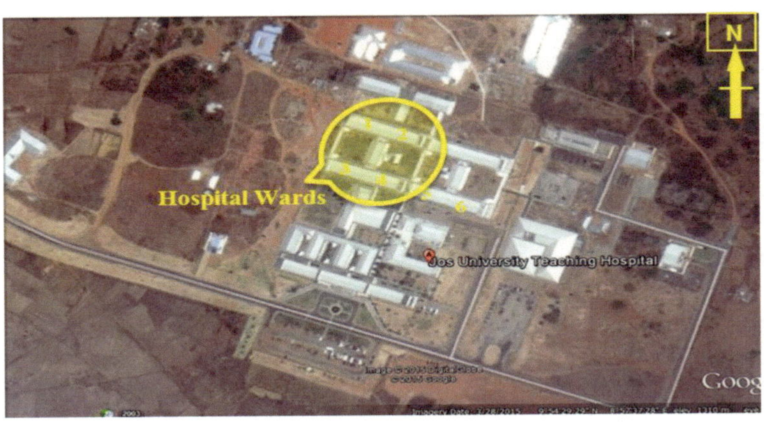

Figure 6.
Site location of Jos University Teaching Hospital, Jos, Plateau State, Nigeria.

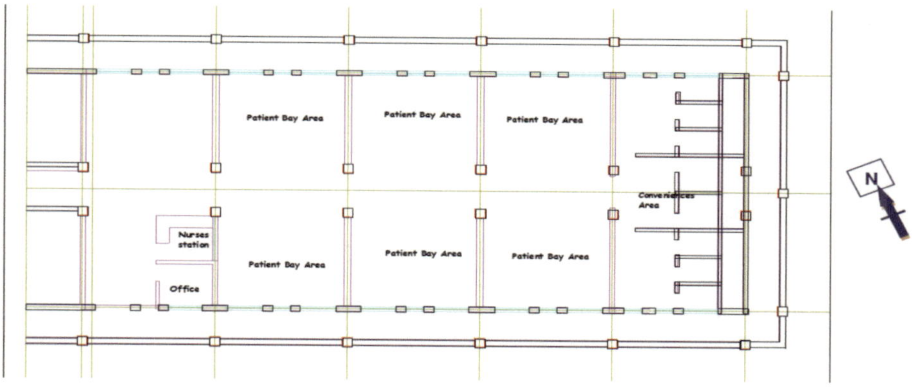

Figure 7.
Typical floor plan of Teaching Hospital ward.

Figure 8.
(a and b) Exterior views of Teaching Hospital complex.

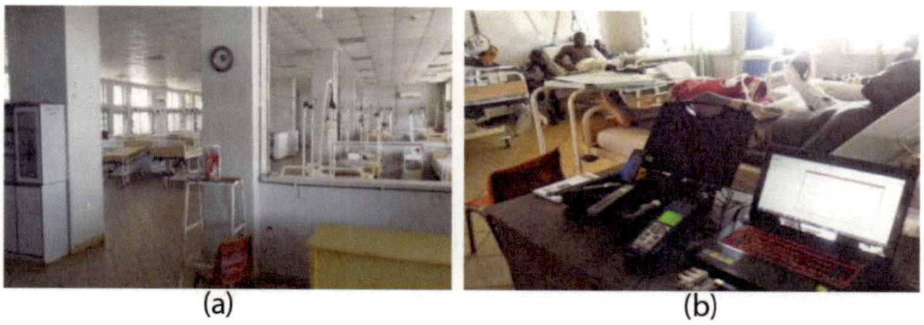

Figure 9.
(a and b) Interior space images of the Teaching Hospital.

supply to the buildings. Artificial mechanical ventilation and lighting is readily available within the ward buildings. The building complex orientation is facing the Northwest and Southeast direction which allows for maximum utilisation of daylighting (**Figure 6**). The Teaching Hospital space organisation is the open or Nightingale type.

The Teaching Hospital wards spatial configurations are the multi-bed bays segmented into three. This provides the nurses with direct observation of patients but at the expense of patients' privacy. The facades and fenestration design of the wards were installed with glazed aluminium panelled windows on both axis facing Northeast and Southwest with a window to wall ratio (WWR) of about 50%. Also provided are top daylight windows for deeper penetration of light into the hospital ward. The glazing on the windows are double pane clear glass used on aluminium panels. **Figures** 7 and **9** shows a sketch of the hospital ward floor plan, views and three dimensional elevations of the hospital wards building.

2.4 Measurement and evaluation

The assessment of IEQ performance is based on the measurement of its four-factor parameters in each of the selected ward buildings in the case study hospitals, which was carried out within 3 months period. The main objective of the physical measurement of IEQ is to investigate the thermal, acoustic, lighting and indoor air conditions in the hospital buildings in order to determine their comfort level. **Table 2** shows a list of IEQ parameters measurement variables. The objective physical measurement is used in cross-validating the IEQ parameters measures in comparison with certain international standards and guidelines. The specific objectives of physical measurement of IEQ parameters were:

a. To measure the thermal condition (temperature and humidity) in the selected hospital wards in order to determine the thermal quality.

b. To determine the acoustic quality by measuring sound level in dBA.

c. To determine the lighting quality based on the measurement of lighting intensity.

d. To measure carbon dioxide and carbon monoxide to determine the indoor air quality.

The procedure for the measurement of the indoor environmental variables is as described in BS EN ISO 28802 [36]. The variables measurement was conducted intermittently at a central location in-between patients' bed in the selected hospital wards. **Table 3** shows an overview of the field measurement of IEQ, and the IEQ variables were measured using the IEQ mobile measurement station as shown in **Figure 9b**. The measuring instrument was set to measure the variables continuously for a period of 2 hours at 5 minutes intervals. The variables were measured at a height of 900 mm above the floor level with the IEQ mobile measurement station positioned within ward building in a central location. The instruments that made up the IEQ mobile measurement station and the specific IEQ variable of measurement is described in detail in **Table 2**. The data logger, which is calibrated both before and after measurement, is set up 1 hour before the commencement of data acquisition on each day of data collection.

IEQ parameters measurements were taken in each of the selected ward buildings indoor environment for three consecutive times over a period of 3 months. This is

IEQ parameter	Measurement variables	Acceptable benchmark for hospitals
Thermal quality	Temperature (°C)	21–24°C [26] 23–26°C [27] 24–33°C Adaptive comfort. BS 15251 [28] 27–37°C [29]
	Relative humidity (%)	30–60% [30]
Acoustic quality	Sound intensity level (dBA)	<40 dBA [31] <60 dBA [32]
Lighting quality	Light Illuminance level (lux)	100–150 lux [33] 100–225 lux [34]
IAQ	Carbon dioxide (ppm) Carbon monoxide (ppm)	<700 ppm [28], [35] <9 ppm [35]

Table 2.
Measurement variables for objective physical measurements of IEQ.

IEQ field measurement	Selected case study hospitals	
	Plateau Specialist Hospital	**Jos University Teaching Hospital**
Building location	A236-Club Road, Jos. 9° 53'42.9"N and 8°53'02.2"E	Lamingo, Jos. 9° 54'27.5"N and 8° 57'37.5"E
Building unit type	Surgical and orthopaedic ward	Surgical and orthopaedic ward
Façade orientation	NW-SE	NE-SW
Design configuration reference [24]	Corridor or continental type	Open or nightingale type
Spatial layout	2-Bed space units of ward rooms	Open—plan multibed units
Measurement variables	Temperature, humidity, carbon dioxide, carbon monoxide, light Intensity, sound level	
Measurement executions	3 Months (April–June)	
Measurement times	2 Days per hospital wards per month (10 am–3 pm)	

Table 3.
An overview of objective field measurement.

Measurements	Boundary conditions
Specialist Hospital (NE-SW) Teaching Hospital (NW-SE)	• All windows open for ventilation within the period of measurement from 10 am to 3 pm. • Window blinds were rolled up to about 60%. • All external doors were closed throughout the period of measurement. ○ Artificial light off. ○ Mechanical ceiling fan off.

Table 4.
Physical measurement boundary conditions.

to ensure that any variation in the measured variables over time is being taken into consideration. The data collected from the different selected ward buildings of each hospital were average for statistical analysis. **Table 4** Shows the boundary conditions applied during the field measurements.

3. Results

This section discusses the relationship between IEQ and hospital wards having different building characteristic (orientation and spatial layout). The IEQ parameters variable data collected during the field measurements are, air temperature, relative humidity, sound intensity level, light intensity level (illuminance), carbon dioxide (CO_2), and carbon monoxide (CO) as shown in **Table 2**. The results of the field measurements of physical indoor environmental variables in the different case study hospitals and taken in three different months is described below. The results of this empirical measurement of the IEQ variables is presented as a descriptive analysis summary for the two different hospitals ward buildings measured in three different months. **Table 5** shows a statistical summary of the monthly measured variables in the case study hospital wards.

Case study	Statistics	Temperature (°C)	Relative humidity (%)	Sound intensity (dBA)	Light intensity (lux)	CO_2 (ppm)	CO (ppm)
Specialist Hospital	Mean	30.89	55.91	71.57	248.01	463.28	6.17
	SD	2.11	5.08	6.88	62.08	58.67	4.01
	Minimum	26.20	43.00	54.10	173.60	400.00	3.00
	Maximum	35.00	63.20	83.80	342.60	608.00	14.00
Teaching Hospital	Mean	30.41	58.47	65.22	333.75	450.01	4.48
	SD	1.80	8.34	7.21	67.54	34.26	1.87
	Minimum	26.80	40.70	52.70	189.20	393.00	2.00
	Maximum	35.00	71.30	75.00	432.00	517.00	10.00

Table 5.
Summary of objective empirical measurements in the hospital wards.

3.1 Thermal quality in hospital wards

The thermal quality in the selected hospital buildings is measured by indoor air temperature and relative humidity. A hospital environment is seen as being traumatic where excessive temperature could have great impact on the building occupants [37]. Temperature is a major determinant of thermal quality which also has relative humidity as its function. The two variables considered for thermal quality, temperature and relative humidity within the indoor spaces of the hospital ward buildings were measured consecutively within a period of 3 months. This section analyses the temperature and relative humidity in the selected hospital ward buildings according to the hospital wards orientations and period of measurement. The hourly average outdoor weather condition recorded within the period of measurement in each of the case study hospital ward is shown in **Table 6**.

3.1.1 Thermal quality in Specialist Hospital (NE-SW orientation)

The average mean outdoor temperatures for the period of measurement are shown in **Table 6**. The indoor and outdoor temperature difference in April and June is significantly different from the difference in May with about 1.5°C. The indoor and outdoor thermal variables measured showed a strong periodic trend. The mean monthly temperature measured in the Specialist Hospital is presented in **Figure 10**. The recorded mean temperature within the period of measurement ranged from 29.9 to 35.3°C. There was a relatively linear reduction in mean indoor temperature from April to May. A maximum temperature range of 33.2–36.2°C was recorded in April while a minimum temperature range of 29.0–31.9°C was recorded in June as presented in **Figure 11**. The variation in temperature increased with time in April while decreasing in May and June. The minimum temperature (30.8°C) was recorded in June at about 2.00 pm hours while the maximum temperature (36.2°C) was recorded at about 10.00 am in April (**Figure 11**). The acceptable limits of temperature shown in **Figures 11** and **12** are based on adaptive thermal comfort.

The relative humidity recorded showed an inverse variation to temperature within the period of measurement in this hospital buildings. The mean relative humidity recorded in April at 56.5% increased to 63.1% in June as a result of the significant effect of annual rainfall on the indoor air temperature. The minimum relative humidity

Period	Variable	10 am	11 am	12 pm	1 pm	2 pm	3 pm	Average
				Specialist Hospital				
April	Temperature (°C)	28.9	29.8	31.7	32.8	33.1	33.3	31.6
	Wind (m/s)	2.2	2.2	1.8	0.9	0.9	0.9	1.5
	Humidity	52.0%	43.0%	36.0%	31.0%	34.3%	32.8%	38.2%
May	Temperature (°C)	27.3	28.9	30.4	31.0	31.3	31.5	30.1
	Wind (m/s)	2.9	2.3	2.3	2.3	2.4	2.4	2.5
	Humidity	74.9%	61.9%	51.8%	44.6%	41.8%	45.1%	53.4%
June	Temperature (°C)	24.6	26.0	26.9	27.9	27.9	28.3	26.9
	Wind (m/s)	3.1	2.5	2.5	2.5	2.6	2.6	2.6
	Humidity	90.0%	74.4%	62.3%	54.6%	50.2%	48.4%	63.3%
				Teaching Hospital				
April	Temperature (°C)	29.4	30.6	31.7	32.2	32.2	32.2	31.4
	Wind (m/s)	2.8	2.2	2.2	2.2	2.3	2.3	2.3
	Humidity	51.0%	43.0%	34.0%	32.0%	30.0%	29.0%	36.5%
May	Temperature (°C)	27.8	28.9	29.9	30.5	30.5	30.5	29.7
	Wind (m/s)	2.3	2.4	1.9	0.9	0.9	0.9	1.6
	Humidity	73.4%	64.6%	49.0%	46.1%	45.7%	43.7%	53.7%
June	Temperature (°C)	25.0	26.5	26.9	27.4	29.5	30.1	27.6
	Wind (m/s)	2.5	2.5	2.0	1.0	1.0	1.0	1.7
	Humidity	88.2%	74.4%	58.8%	55.4%	51.9%	50.2%	63.1%

Table 6.
Field measurement of hourly average outdoor weather condition.

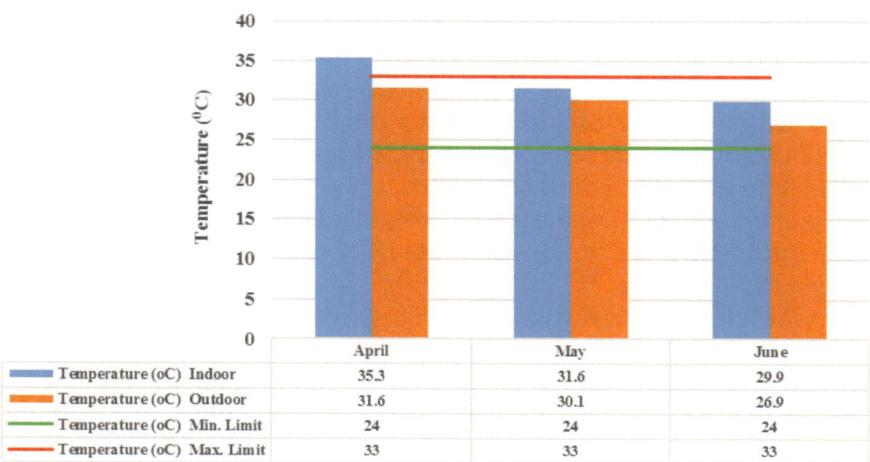

Figure 10.
Mean relative humidity in Specialist Hospital wards.

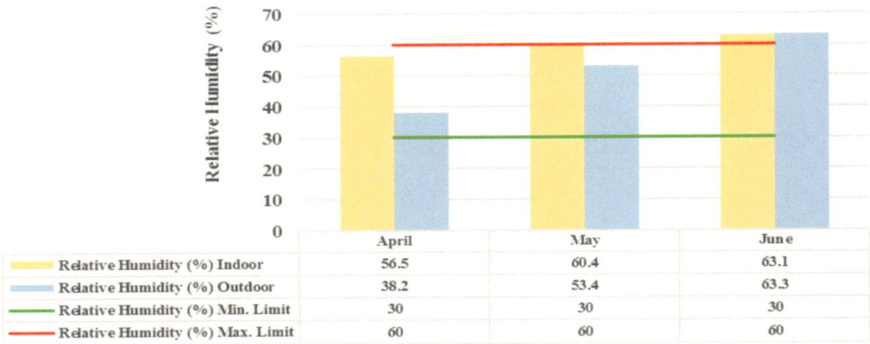

Figure 11.
Variation in temperature in Specialist Hospital wards.

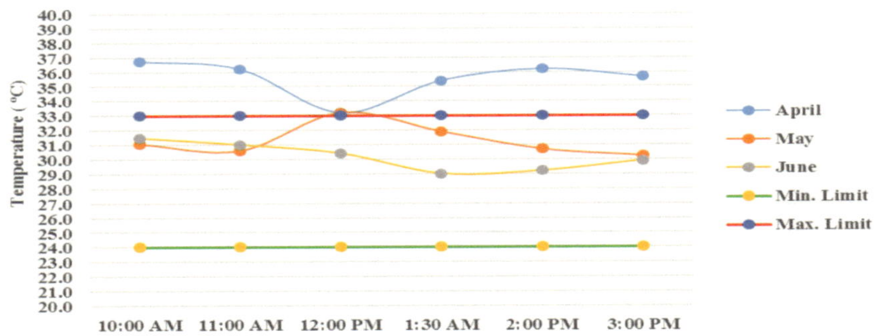

Figure 12.
Mean temperature in Specialist Hospital wards.

recorded within the period of monitoring is between 54.9 and 57.7%, while the maximum is between 60.8 and 64.8%. The relative humidity are higher within the morning hour and lower in the afternoons. **Figure 13** shows variations in the relative humidity in the Specialist Hospital which is relatively uniform in May and June.

3.1.2 Thermal quality in the Teaching Hospital (NW-SE orientation)

The variation trend of temperature is the same as the other two case study hospitals, as there was also a temperature decrease recorded for May and June. The mean temperature and relative humidity recorded during the monitoring periods is shown in **Figures 14** and **15**. The mean temperature range recorded in each of the ward buildings is between 29.3°C recorded in May and 32.9°C recorded in April. The mean temperature in May and June which are almost invariable, are relatively lower than the mean temperature recorded in April. Temperature variation as shown in **Figure 16** decreases with about 4.0°C within an hour and increased again with the same magnitude within 30 minutes both in May and June. The temperature was steady in April with only a variation at about 12.00 noon where it recorded its highest temperature of 35.4°C.

The mean relative humidity level ranged between 56.9 and 65.8% (**Figure 15**) as recorded in the ward buildings. There is no proportionality in the relationship between relative humidity and temperature in this hospital. As with the other hospital buildings, there are differences in the relative humidity measured at different periods. The variation of relative humidity as recorded at different periods is as shown in **Figure 17**. In April, the average relative humidity ranged from 53.1 to 69.2%, which was quite higher than the relative humidity recorded for May and June. The month of May recorded the least steady variation in relative humidity

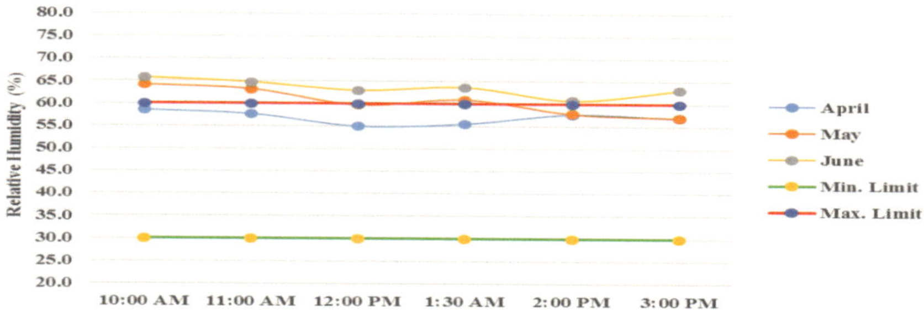

Figure 13.
Variation in relative humidity in Specialist Hospital.

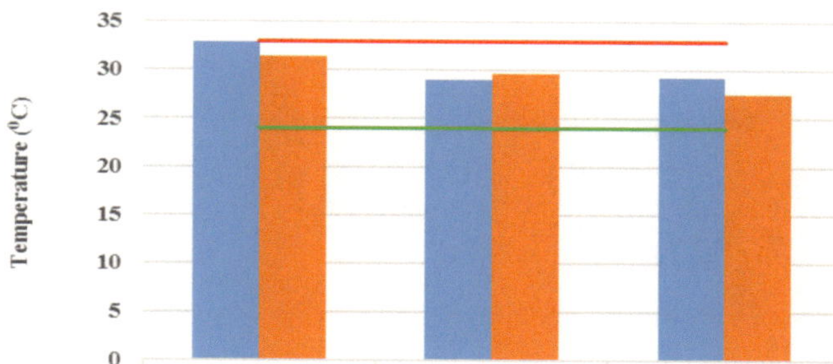

	April	May	June
Temperature (oC) Indoor	32.9	29	29.3
Temperature (oC) Outdoor	31.4	29.7	27.6
Temperature (oC) Min. Limit	24	24	24
Temperature (oC) Max. Limit	33	33	33

Figure 14.
Mean temperature in Teaching Hospital wards.

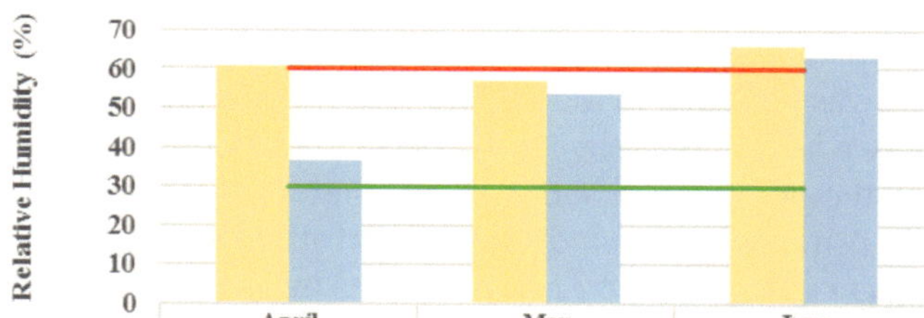

	April	May	June
Relative Humidity (%) Indoor	60.8	56.9	65.8
Relative Humidity (%) Outdoor	36.5	53.7	63.1
Relative Humidity (%) Min. Limit	30	30	30
Relative Humidity (%) Max. Limit	60	60	60

Figure 15.
Mean relative humidity in Teaching Hospital wards.

having a range difference 6.9% as compared to April and June having a range difference of 10 and 16.2% respectively.

3.1.3 Thermal quality variations by hospital ward buildings

The monitoring of indoor temperature and relative humidity was carried out with the IEQ mobile measurement station data logger positioned within the ward buildings in each of the case study hospital wards. The thermal qualities in each of these hospital wards differ since their design, configurations and orientation also differs. The indoor temperature and relative humidity therefore varied according

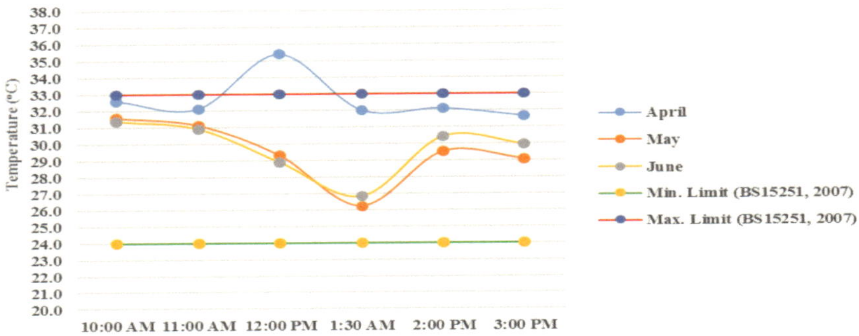

Figure 16.
Variation in temperature in Teaching Hospital wards.

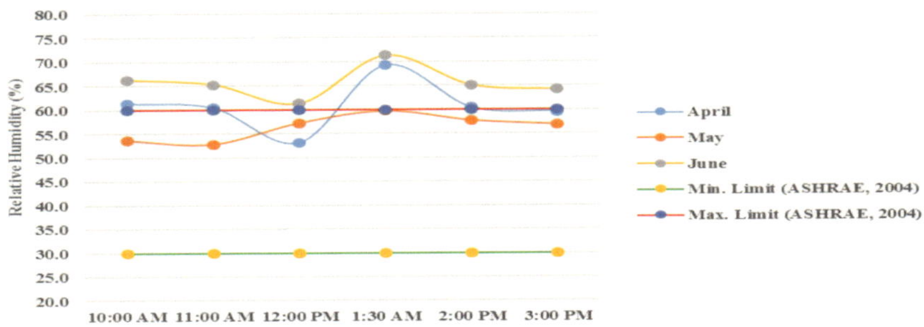

Figure 17.
Variation in relative humidity in Teaching Hospital wards.

to the hospital ward orientation. The variation in temperature in each hospital measured within a given period is shown in **Figure 18** was highest (36.7°C) in the Specialist Hospital wards as measured in April while the lowest temperature (26.2°C) was recorded in May in the Teaching Hospital. The indoor temperature of both two hospitals changes with period of monitoring. The variations in the monthly temperatures in both hospital buildings can be said to be having the same trend. In April, the mean indoor temperature as measured in the hospital ward buildings ranged between 33.2 and 36.7°C in the Specialist Hospital, while the mean temperature ranged between 32.0 and 35.4°C in the Teaching Hospital. The same trend is evident in measurement for May and June for both hospitals; however, there was a drop in temperature of between 1.5 and 4.5°C in May and June respectively. There was no any particular trend in temperature variations at specific time

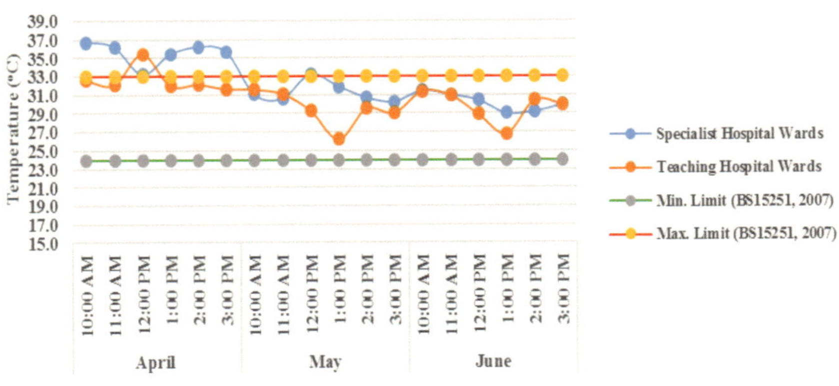

Figure 18.
Monthly temperature variation in the hospital ward buildings.

as measured in each hospital. Between the hours of 11.00 am and 12.00 noon in April, the temperatures increased from 32.1 to 35.4°C in the Teaching Hospital while decreasing from 36.2 to 33.2°C in the Specialist Hospital.

The indoor temperature levels in both hospital wards were influenced by incoming sunlight. The orientation of the Teaching Hospital wards is such that the façade windows are exposed to direct sunlight both at sunrise and sunset. However, the corridor provided on the Southwest façade provides shading against direct penetration of solar radiation. Due to heat gains from sunlight, the temperature in the Teaching Hospital wards were higher between the hours of 10.00 am and 12.00 noon than the hours between 1.00 pm and 3.00 pm. Temperatures in the Specialist Hospital wards were generally higher through the period of measurement as compared to temperatures in the Teaching Hospital wards.

The variation trend of temperature in both Specialist Hospital wards tend to reduce between the hours of 1.00 noon and 3.00 pm in May and June. As much as the Teaching Hospital has the minimum temperature range recorded within the periods of measurement, the fluctuations in the temperature within specific time of the day is highest. Also, the mean relative humidity recorded lowest and highest values both in the Teaching Hospital wards in the month of June (**Figure 19**). Relative humidity tends to increase from April to June as measured in the Specialist Hospital and Teaching Hospital. On an average, lower relative humidity was recorded at 12.00 noon in each monitoring day of the hospital wards.

The thermal quality of any building occupant depends more on temperature as the most important indoor environmental variable. Higher temperatures were recorded in the Specialist Hospital wards for the period of measurement as compared to the Teaching Hospital wards. The Specialist Hospital wards and Teaching Hospital wards are both located on the highland of Plateau. The average outdoor temperature difference between the locations of the two hospital ward buildings was below 1°C within the period of measurement. The mean temperatures as measured in the two hospital ward buildings ranged from 30.4 to 34.9°C. The relative humidity level on the other hand was from 55.9 to 58.5%. According to international standards and guidelines, the mean indoor temperature recorded in both case study hospital wards are above acceptable limits of 23–26°C for occupants' comfort [28]. However, the temperature range recorded in the Teaching Hospital wards are within acceptable limits of adaptive comfort (24–33°C) stated in BS 15251 [28].

The Specialist Hospital wards whose average temperature recorded the highest (35.3°C) in the month of April still fall within the acceptable limit of 27–37°C as opined Nicol and Humphreys [29]. The average monthly indoor temperature pattern was relatively stable in the Teaching Hospital wards with open-plan configuration and windows that allows for cross ventilation than in the Specialist Hospital wards.

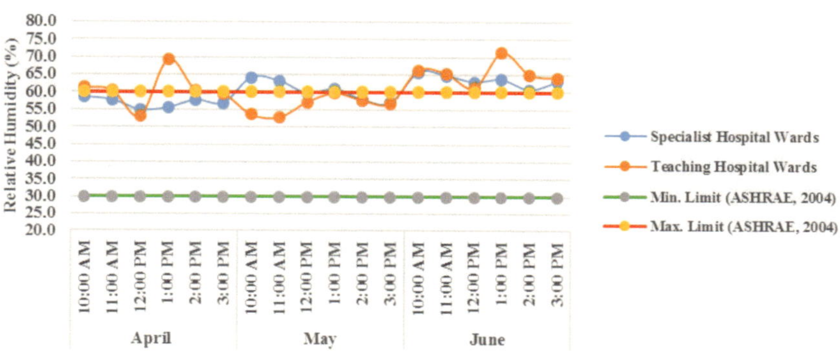

Figure 19.
Monthly relative humidity variation in the hospital ward buildings.

The indoor relative humidity levels in both hospital ward buildings fall within acceptable range of (30–60%) as provided for in international standards [30]. The mean relative humidity level for April in all hospitals is lower, as April always mark the end of dry season. According to Environmental Protection Agency [38], high humidity level in buildings stimulates the breeding of micro-organisms which have adverse effect on building occupants especially in healthcare facilities. However based on [30] which recommended that relative humidity should not be greater than 65%, the mean relative humidity can said to be within acceptable range. The indoor humidity level in both hospital wards were not the same for the period of measurement. The Teaching Hospital wards recorded humidity levels which were significantly higher than the outdoor humidity as compared to the Specialist Hospital wards. This was strongly influenced by the wide windows provided on adjacent façade walls in response to variations in outdoor humidity.

The temperatures and relative humidity levels recorded in the hospitals are not uncommon for naturally ventilated buildings located in the tropical regions of the world. The high temperatures recorded in both hospital wards might be due to the exposure of their facades to direct solar radiation. The provision of appropriate shading through landscape elements could help in preventing overheating within the hospital wards. Hospital wards design in the tropics should there incorporate shading design principles either passive or active towards achieving thermal balance in the buildings. The design configuration of the Teaching Hospital wards having cross ventilation through the provision of wider windows allows for the free movement of air in and out of the wards. The free flow of air tends to cool the heated air within the wards thereby reducing the indoor temperature level. The Specialist Hospital wards on the other hand have closed-plan design configuration with the rooms not having proper ventilation that is required to provide thermal balance for the patients.

3.2 Acoustic quality in hospital wards

3.2.1 Acoustic quality in Specialist Hospital (NE-SW orientation)

The measurement of background noise level within the hospital wards were carried out between the hours of 10.00 am and 3.00 pm. This have excluded the influence of sound level due to the activities of visitors whose permitted visiting hours is between 3.30 pm and 5.00 pm. The mean sound intensity levels in the Specialist Hospital is highest in May and lowest in June which range between 65.5 and 71.8 dBA. The mean difference in sound intensity level between May and June is higher than the mean difference between April and May. The mean sound intensity levels in the Specialist Hospital is presented in **Figure 20**. The variations in sound intensity levels tend to reduce in May and June while increasing in April from 12.00 noon to 2.00 pm as shown in **Figure 21**. The highest sound intensity level of about 78.2 dBA was recorded at 12.00 noon in May and the minimum of 56.9 dBA was recorded at 1.30 pm in June. The variation in sound intensity level within the period of monitoring was affected mainly by the noise level resulting from the number of visitors found within the ward buildings.

3.2.2 Acoustic quality in Teaching Hospital (NW-SE orientation)

The mean difference in sound intensity level measured within the Teaching Hospital as shown in **Figure 22** is 8.6 dBA. Just like the Specialist Hospital, the highest sound intensity level was recorded in May and the lowest in June. The sound intensity level as measured within 2 hours were higher in May but lower in June.

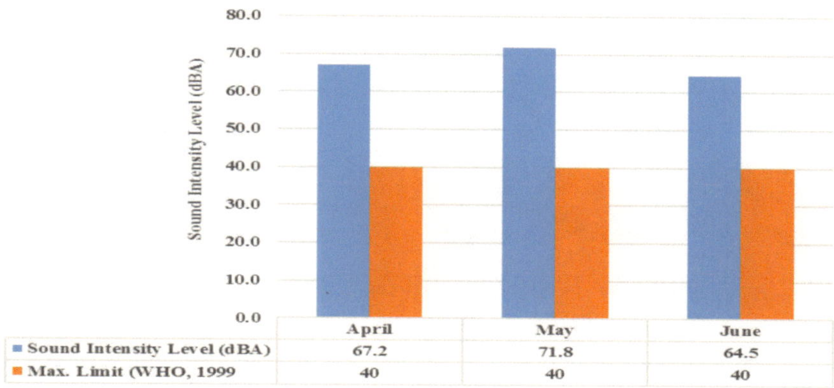

Figure 20.
Mean sound intensity level in Specialist Hospital wards.

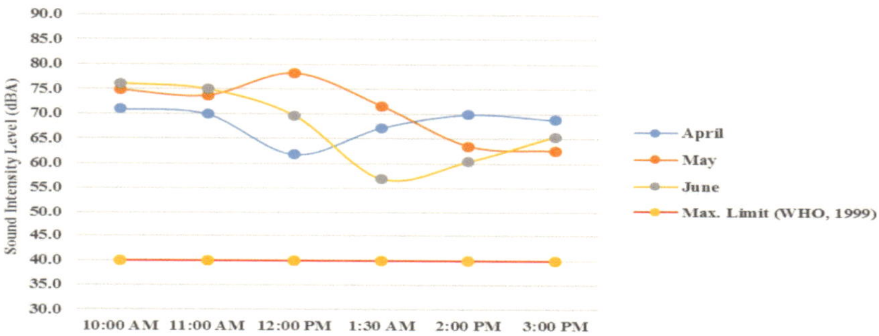

Figure 21.
Variation in sound intensity level in Specialist Hospital wards.

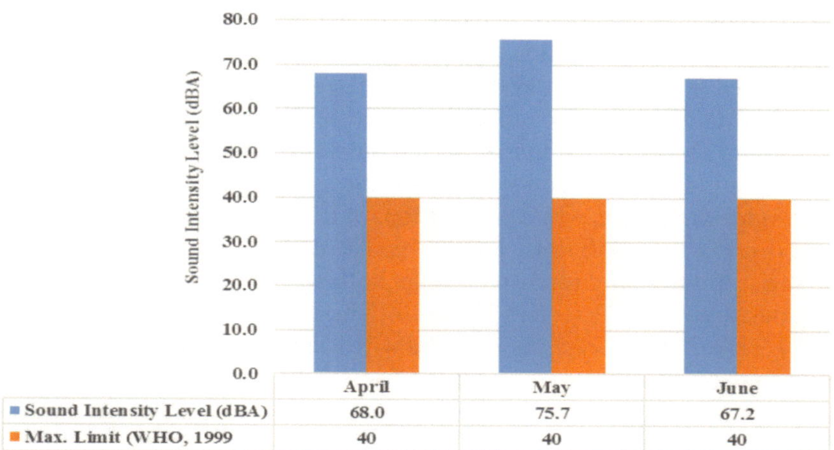

Figure 22.
Mean sound intensity level in Teaching Hospital wards.

The highest sound level was recorded at 10.00 am (85.1 dBA) in May while the lowest was recorded at 1.00 pm in June. There was no particular trend in the monthly variations of sound intensity level, as the sound level increased and decreased alternately in April but, reversed is the case in May. However in June, the sound intensity level decreased from 73.6 dBA at 10.00 am to 59.9 dBA at 1.00 pm. But there was an increase also to 70.0 dBA at 2.00 pm (**Figure 23**).

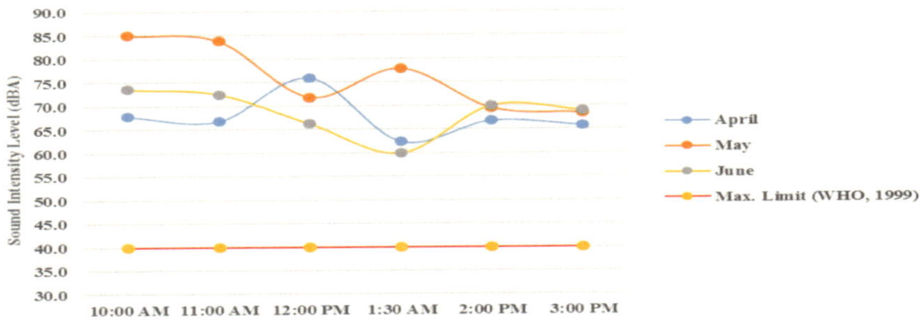

Figure 23.
Variation in sound intensity level in Teaching Hospital wards.

3.2.3 Acoustic quality variations by hospital ward buildings

Acoustic quality is another important element that needs proper consideration in the design of hospital wards environment. Apart from the fact that it affects sleep, it also contribute significantly to patient's healing process. The world health organisation [31] have defined an acceptable maximum limit of sound intensity level range of 40 dBA. **Figure 24** shows the monthly variations in sound intensity levels in the two case study hospitals. The level of variations of sound intensity with time almost took the same pattern. The mean differences in sound intensity levels recorded ranged between 8.1 and 18.6 dBA in the Specialist Hospital, and 12.6 and 14.4 dBA in the Teaching Hospital. Both the minimum and maximum mean differences in sound intensity level were recorded in Specialist Hospital wards. This is an indication that, there was a wider variation in the sound intensity levels as measured in the Specialist Hospital wards than it was in Teaching Hospital wards. The highest sound intensity level (83.8 dBA) was recorded in the Teaching Hospital at 11.00 am in May while, the lowest (56.9 dBA) was recorded in the Specialist Hospital wards also at 1.00 pm but in June.

The variation in sound intensity levels is smaller in the Specialist Hospital wards with SD = 6.88 while the variation is more in the Teaching Hospital with SD = 7.21 (**Table 5**). The sound intensity level is highest in the Specialist Hospital (71.6 dBA) whose background noise is also influenced by vehicular traffic flow aside noise created by staff and visitors activities within the ward buildings. The location of the Specialist Hospital ward buildings are adjacent to a major road within the city centre with high vehicular traffic, while the locations of the Teaching Hospital is away from vehicular disturbances as the only is the vehicular access leading to the hospital.

The variations in sound intensity level in both hospitals ward buildings almost followed the same trend as seen in **Figure 24**. The average variation in sound intensity level is higher in the Teaching Hospital with a standard deviation of about 0.4 greater than the standard deviation for the Specialist Hospital. **Table 5** shows the mean sound intensity levels recorded in each hospital wards within the period of measurement. The sound intensity level can be said to be relatively the same in both two hospitals within the 3 months measurement periods having a maximum difference in intensity level of less than 4 dBA. The indoor sound levels in the hospitals are all above 40 dBA which is higher than the acceptable range of 30–40 dBA [31]. The highest sound intensity level for each of the hospitals is recorded in May which is slightly above 70 dBA.

A background noise level of up to 35 dBA is considered by the World Health Organisation (WHO) as acceptable in patient's wards during the day, and a sound level of not greater than 40 dBA at day and night peak to allow for patients' rest. Measures towards reducing background noise into the hospital wards is required

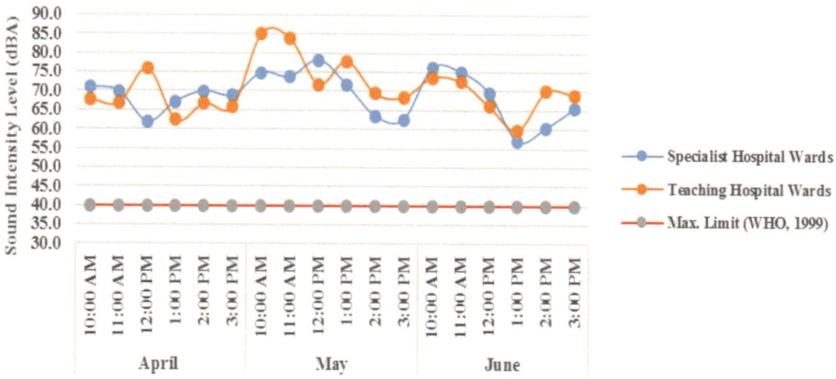

Figure 24.
Monthly sound intensity level variation in the hospital ward buildings.

in both case study hospitals as sound intensity levels recorded within the period of measurement ranged from 56 to 85.1 dBA. This will go a long way in mitigating such impact as sleep loss and emotional exhaustion that could hamper the healing processes. The background noise affecting the Specialist Hospital wards might have been influenced also by vehicular traffic due to the location of the hospital in an area within the city centre, bounded by major roads. Other sources of sound that is typical includes; patients' reaction to their health situation through groaning or crying, medical equipment and patient's interaction with their caregivers. One major way of reducing such sound effect is by masking, which is beyond the control of the design team. The use of sound absorptive ceiling tiles having minimum NRC of between 0.90 and 0.95 [39] can be implemented in achieving high acoustics benefits in the patient wards. Where the acoustic condition of hospital wards is improved through design and selection of material components, it will improve on the patient healing process and reduce length of stay.

3.3 Lighting quality in hospital wards

3.3.1 Lighting quality in Specialist Hospital (NE-SW orientation)

The mean illuminance level recorded in the Specialist Hospital fell above 100 lux as the minimum recommended for hospital wards [33, 34]. The mean difference in illuminance level between different periods of measurement ranged between 21.4 and 70.5 lux. **Figure 25** shows the mean illuminance levels recorded in the Specialist Hospital. The variation in light intensity level with time showed an increased in April but a decrease in both May and June from 12.00 am to 2.00 pm. Reverse is the case as from 2.00 pm to 12.30 pm (**Figure 26**). Higher illuminance level was recorded in April between the hours of 12.00 noon and 1.00 pm.

3.3.2 Lighting quality in Teaching Hospital (NW-SE orientation)

The mean illuminance levels recorded lower in May which was less than 200 lux. The mean illuminance levels in May and June are near equal in intensity and fall within the range of 184.6–339.6 lux. There was a mean difference in intensity level of about 101 lux between both May and June, and April. **Figure 27** shows the mean intensity level recorded for each measurement period. As also shown in **Figure 28**, the periodic increase and decrease in illuminance level pattern is almost similar, however, the change is higher in April than in May and June.

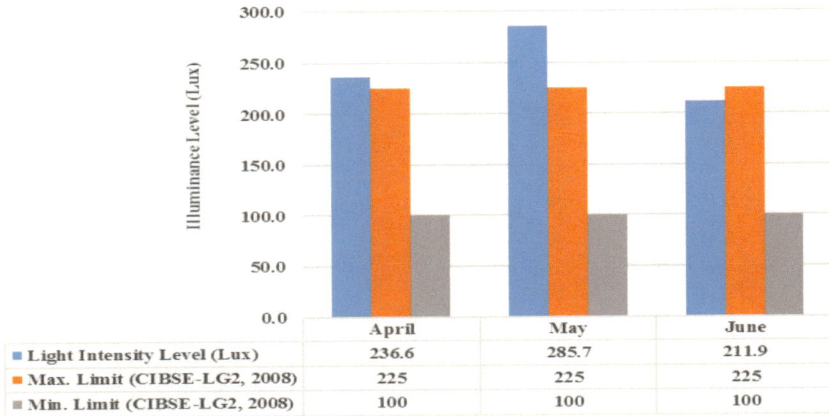

Figure 25.
Mean light intensity level in Specialist Hospital wards.

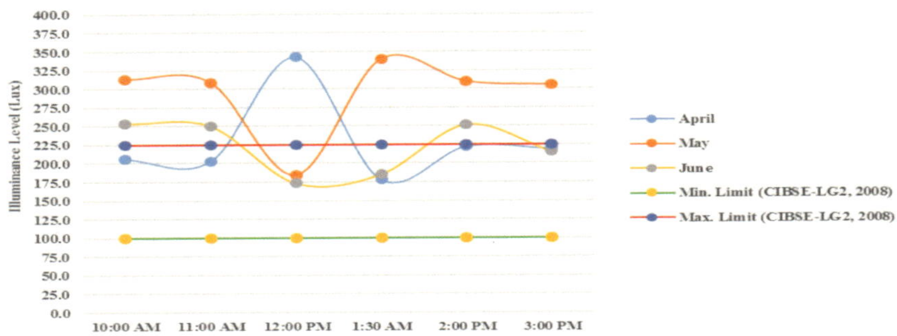

Figure 26.
Variation in light intensity level in Specialist Hospital wards.

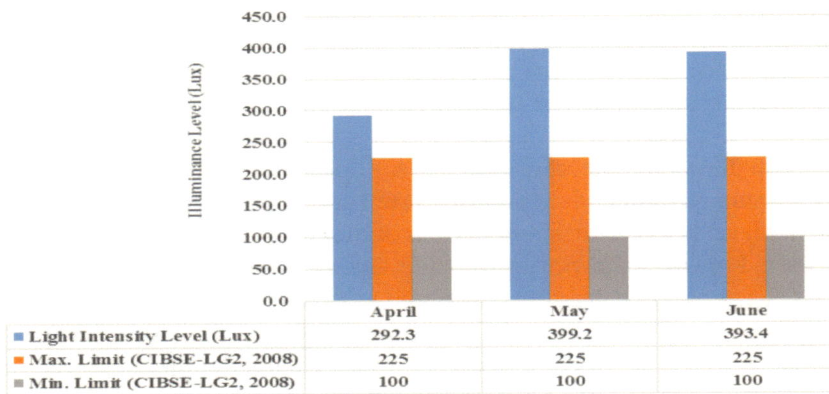

Figure 27.
Mean light intensity level in Teaching Hospital wards.

3.3.3 Lighting quality variations by hospital ward buildings

The monthly variations in light intensity levels in the studied hospital buildings is shown in **Figure 29**. The illuminance levels recorded highest in April in the Specialist Hospital wards, while in May and June, it recorded highest in the Teaching Hospital. Generally, the lighting quality in the Specialist Hospital is poorer as compared with what was obtainable in the Teaching Hospital. The

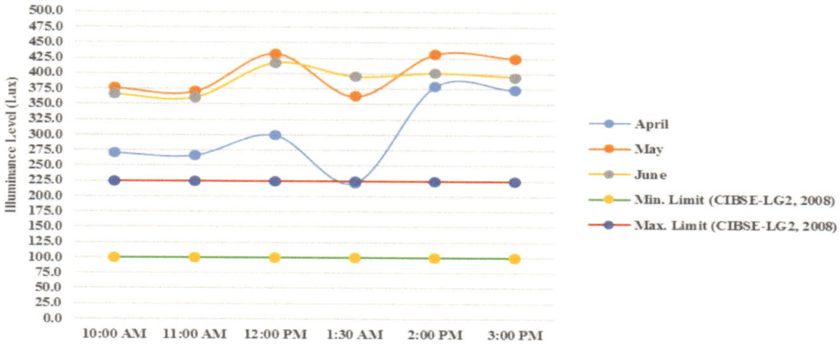

Figure 28.
Variation in light intensity level in Teaching Hospital wards.

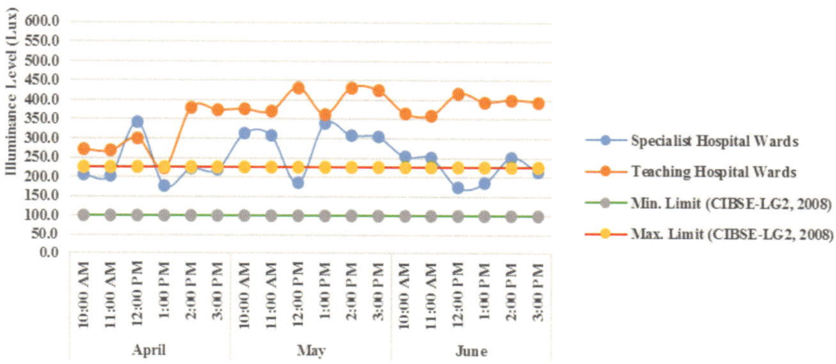

Figure 29.
Monthly light intensity level variation in the hospital ward buildings.

North-East and South-West window facing orientation of the Teaching Hospital allowed for optimum daylighting into the ward buildings. The orientation of the Specialist Hospital wards followed the North-West and South-East direction. Daylight penetration is only through one façade of the ward buildings in the Specialist Hospital where the light intensity level is influenced by sun path position.

The least light intensity level (173.6 lux) was recorded in the Teaching Hospital in the month of May which could have resulted from the effect of cloud cover at the time of measurement. The variation in the lighting quality with time in both hospital ward buildings can said to be relatively equal as their differences in standard deviation is not more than 5.5. Their standard deviation values ranged between 62.1 and 67.5 whose difference cannot be said to be significant (**Table 5**).

The lighting intensity measured in both hospital ward buildings is an indication that the data were influenced by the variations in the sun direction. The light intensity in the Teaching Hospital wards was relatively higher because of the window size design which was quite large in size. The minimum average illuminance (292.3 lux) recorded in the Teaching Hospital wards was significantly higher than the acceptable limit of 225 lux recommended by CIBSE-LG2 [34]. These higher daylight illuminance level is a result of the hospital wards having window to wall ratio (WWR) of 55% which is even higher than the optimum (25%) recommended by Zain-Ahmed et al. [25] for passive design of windows for daylighting in the tropics.

3.4 Indoor air quality (IAQ) in hospital wards

3.4.1 Indoor air quality (IAQ) in Specialist Hospital (NE-SW orientation)

The mean concentration level of CO_2 is higher in April as compared to the concentration levels in May and June (**Figure 30**). On the other hand, the CO concentration level is higher in June (**Figure 31**). The CO_2 concentration level which all fall within acceptable limits ranged between 439.3 and 608.0 ppm in April, between 399 and 442 ppm in May, and 393 and 455 ppm in June (**Figure 32**). Likewise, the CO concentration level ranged from 4 to 9 ppm in April, from 4 to 8 ppm in May and from 3 to 14.2 ppm in June. The CO concentration level for June as shown in **Figure 33** indicated that only the recorded value at 1.00 pm fall within the acceptable limit that promotes occupants health and comfort.

3.4.2 Indoor air quality (IAQ) in Teaching Hospital (NW-SE orientation)

The results of the measured CO_2 and CO concentration levels in the Teaching Hospital is shown in **Figures 34** and **35**. There was a small variation in the mean concentration levels of both CO_2 and CO in the Teaching Hospital, having a maximum mean difference of 30.3 and 2.3 ppm respectively. The concentration level of CO_2 ranged between 406.8 and 481.0 ppm in April, 410.0 and 492.0 ppm in May, ad 394.0 and 457.8 ppm in June (**Figure 36**). There was quite stability in the concentration levels of CO_2 which were lower than the maximum acceptable range.

CO level ranged from 2 to 5 ppm, 3–10.2 ppm, and 3–6 ppm in April, May and June respectively as shown in **Figure 37**. The highest CO concentration level of 10 ppm was recorded at 10.00 am in May while the lowest concentration level of 2 ppm was recorded between 12.00 noon and 1.00 pm in April.

3.4.3 Indoor air quality (IAQ) variation by hospital ward buildings

Figures 38 and **39** shows the variations in IAQ in the hospital buildings. There was no much variation in CO_2 concentration levels in the hospital buildings within the measurement periods. The CO_2 concentration levels in the in the Specialist Hospital wards ranged between 393 and 608 ppm, while in the Teaching Hospital wards ranged between 394.0 and 492.3 ppm. In April, the mean CO_2 concentration

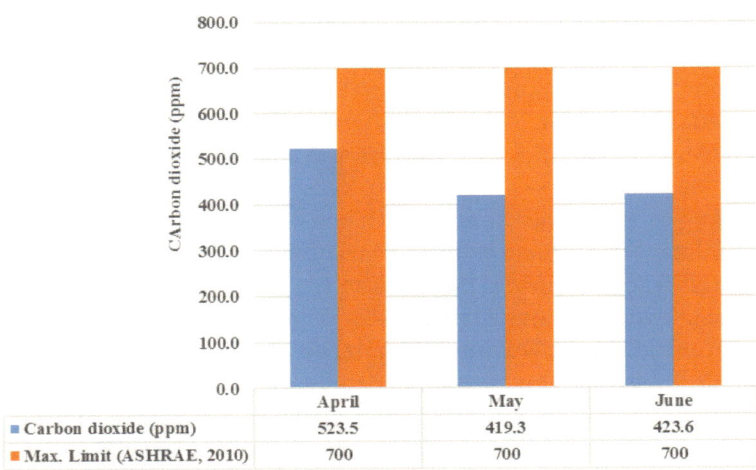

Figure 30.
Mean carbon dioxide concentration level in Specialist Hospital wards.

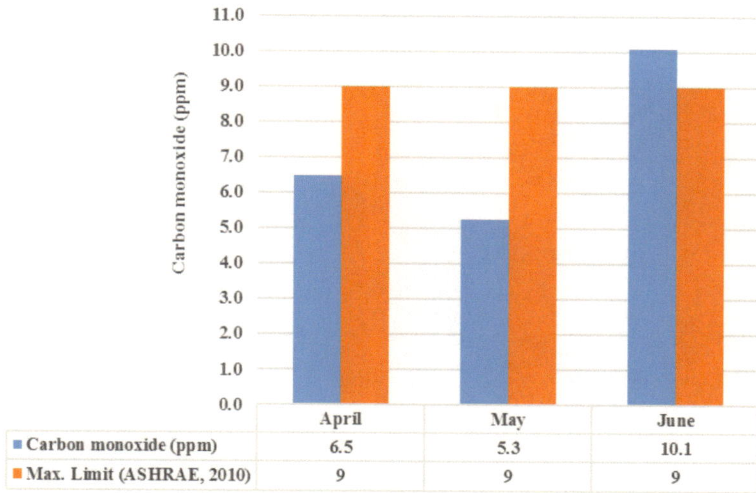

Figure 31.
Mean carbon monoxide concentration level in Specialist Hospital wards.

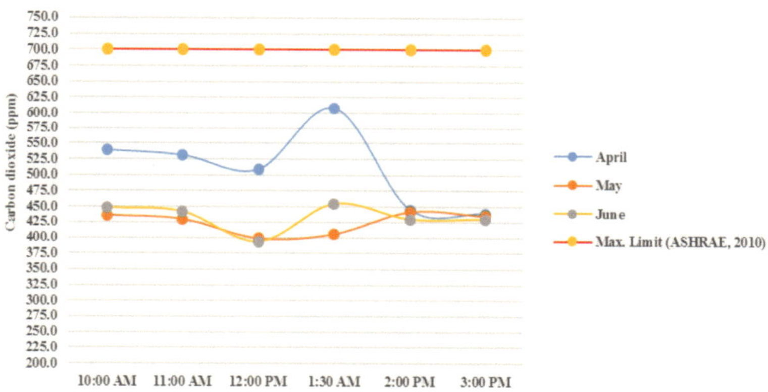

Figure 32.
Variation in carbon dioxide level in Specialist Hospital wards.

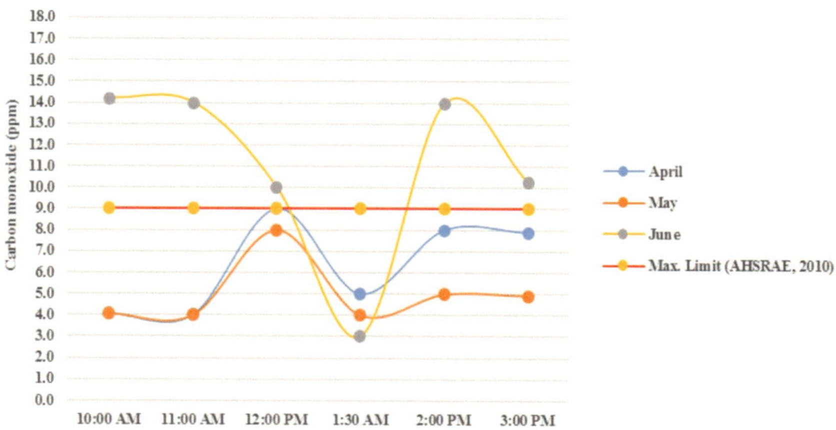

Figure 33.
Variation in carbon monoxide level in Specialist Hospital wards.

level was 408 ppm, 523.5 ppm, and 453.3 ppm for the Specialist and Teaching Hospitals respectively. The highest level of concentration of CO_2 (608 ppm) was recorded in April in the Specialist Hospital and the lowest concentration level

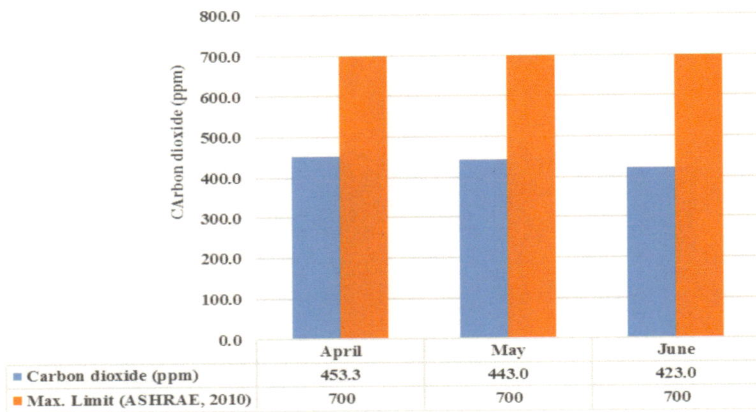

Figure 34.
Mean carbon dioxide concentration level in Teaching Hospital wards.

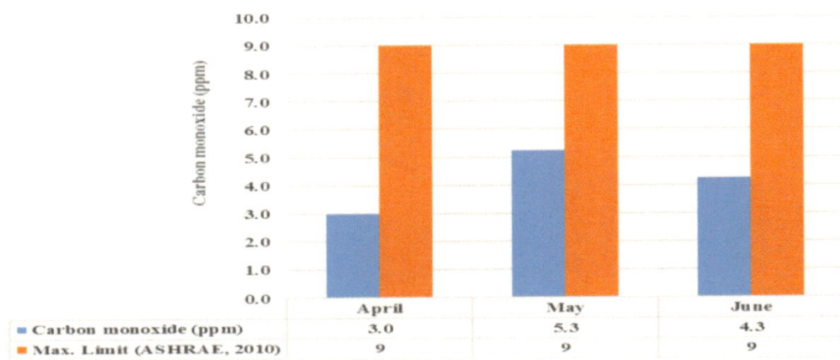

Figure 35.
Mean carbon monoxide concentration level in Teaching Hospital wards.

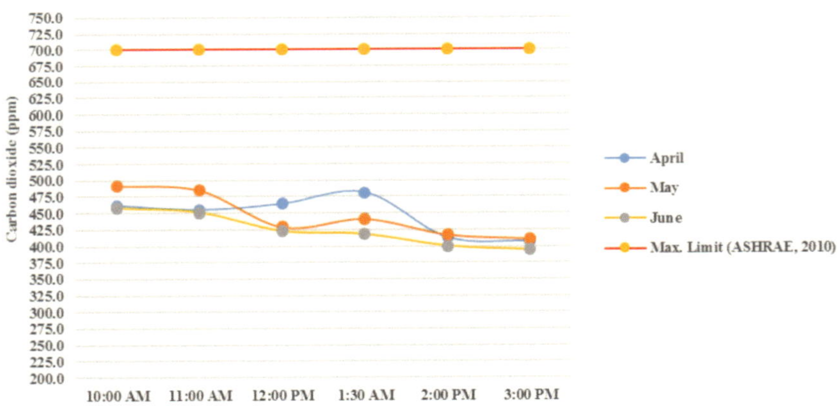

Figure 36.
Variation in carbon dioxide level in Teaching Hospital wards.

(393 ppm) was also recorded in the Specialist Hospital but in June as seen in **Figure 38.** On a general note, the mean concentration levels of CO_2 in both case study hospital wards fall below the maximum recommended limit of 700 ppm [32, 40] within the periods of measurement.

The IAQ in the Teaching Hospital wards is better compared to the Specialist Hospital wards because of its design and age which are contributing factors to the concentration levels of air pollutants. **Figure 39** shows the variations in CO levels

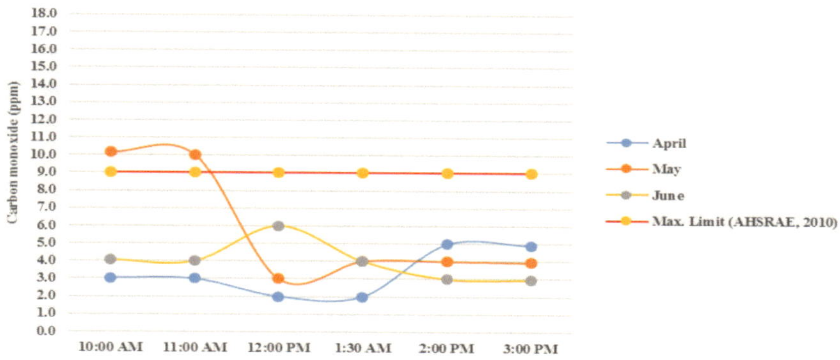

Figure 37.
Variation in carbon monoxide level in Teaching Hospital wards.

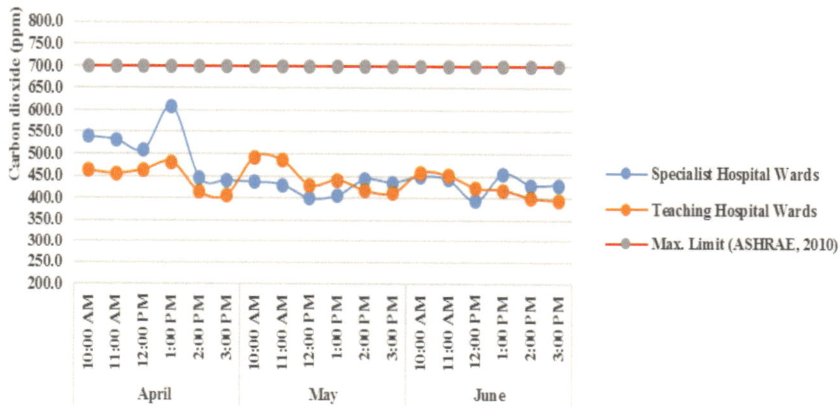

Figure 38.
Monthly carbon dioxide concentration level variation in the hospital ward buildings.

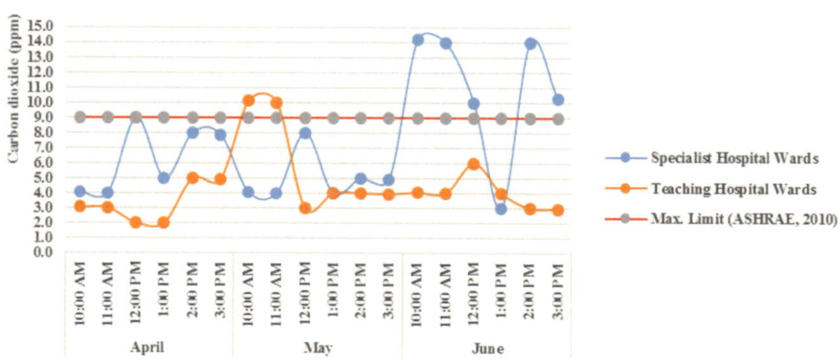

Figure 39.
Monthly carbon monoxide concentration level variation in the hospital ward buildings.

for both hospitals. The maximum recorded CO concentration level for the Specialist Hospital is 14.2 and 10.2 ppm in the Teaching Hospital. All the maximum recorded CO concentration levels are greater than the acceptable limit. The minimum CO concentration level was recorded in the Teaching Hospital. There was generally higher concentration of CO level in June than in April as recorded in both the two case study hospitals. This could be related to increase in humidity as a result of increased amount of rainfall.

The IAQ can be said to be much better in the Teaching Hospital whose design and configuration allowed for free flow of ventilation in and out of the hospital wards. The design of the Specialist Hospital provided no cross ventilation within the ward buildings which could have been the result of the higher level of CO concentration recorded. CO_2 concentration level based on mean estimates within the period of measurement is highest in the Specialist Hospital wards, having a difference in concentration as compared to the Teaching Hospital wards greater than 40 ppm respectively.

4. Discussions

The two hospital ward buildings are naturally ventilated through façade fenestrations with the exception of the Teaching Hospital whose ward buildings have supplemental split-level air-conditioning system. However, this air-conditioning system was not in operation within the periods of monitoring and data collection. The design of windows in the Teaching Hospital allows for proper cross ventilation and air circulation while the Specialist Hospital building design lacks cross ventilation. Based on the design guidelines and standards, the temperatures recorded for the two case study hospitals exceeded the recommended range of 23–26°C [28, 41], though, a temperature range of between 27 and 37°C can provide for occupants' comfort in a building based on human physiological adaptive mechanism [29]. On the contrary, BS EN 15251 [28] provided an acceptable temperature range of 24–33°C for adaptive comfort. The above findings suggests that the increase in the thermal level especially in the Teaching Hospital wards was a result of heat gains from sunlight. Both study hospital wards have no external shading to reduce the effect of heat gain through solar radiation.

The sound intensity level measured in the ward buildings of the two case study hospitals were all above 40 dBA which is above recommended ranges by the world health organisation WHO [31]. The sound intensity levels were high in the month of May than in April and June. Greater variations in sound level was recorded more in the Teaching Hospital wards which recorded higher sound intensity in each of the ward buildings than the Specialist Hospital wards. This might be as a result of more nursing activities related to patient care in the Teaching Hospital than the Specialist Hospital. Furthermore, both staff and caregiver activities in and around the ward buildings in the Teaching Hospital is higher. Sound has been ascertained to have much impact on work performance of building occupants [42] while in hospital environment, it can cause irritation, discomfort and retards patient's healing process [43]. Therefore, acoustic quality should be given much consideration in the design of buildings towards promoting occupants' performance and wellbeing.

One of the basic design indicators for green architecture in creating lighting quality in buildings is daylighting [44]. Natural daylight from the sun when effectively harnessed into a building design can provide a better environment for living and work. Daylighting quality in a building can be influenced by fenestration design, sun path, cloud cover and adjacent physical environmental elements. Building orientation also plays an important role in determining the amount of daylighting in a building as seen from this study. The mean daylight intensity in the Teaching Hospital wards is more than the intensity recorded in the Specialist Hospital wards. The Teaching Hospital ward buildings have their facades and fenestration facing North-east and South-west mostly within the Sun path position. On the other hand, the Specialist Hospital orientation is on North-west and South-east facing where its exposure to the Sun path direction is limited. The Specialist Hospital therefore, has shown the worst lighting quality within the indoor spaces. The mean illuminance level in the Teaching Hospital wards ranged from 292.2 to 397.4 lux and are greater than the 150 lux as the minimum lighting required for hospital wards [33, 45].

For proper medication administration to patients and staff record keeping, it is required that the minimum light intensity level in a hospital ward building should range between 100 and 300 lux [46]. The daily variations in light intensity level in the Teaching Hospital is relatively smaller compared to the variations in the Specialist Hospital. The average light intensity level recorded in the two case study hospitals is an indication that, the application of daylighting features into the design of hospital ward buildings would lead to energy savings and environmental sustainability in healthcare facilities. The results of the measured illuminance level in the hospital wards have shown that ward building orientations have substantial influence in facilitating the use of natural daylighting in hospital buildings. The Teaching Hospital wards with Northwest-Southeast orientation recorded higher lighting intensity throughout the period of measurement than the illuminance level recorded in the Specialist Hospital wards with Northeast-Southwest orientation. However, the lack of shading on the windows and façade walls could pose some lighting challenges to the patients. Nevertheless, this challenges are not within the context of this study.

The maximum allowable threshold limit for CO_2 within an indoor environment given by different international standards is 700 ppm [28, 35, 47]. The concentration level of CO_2 in the ward buildings of the two hospital ward buildings was below 700 ppm within the period of measurement. The highest concentration of CO_2 was recorded in April in all two hospital wards with the Specialist Hospital having the highest. The month of April also recorded the highest temperature in both case study hospitals which indicated that, there is a relationship between CO_2 concentration and temperature. Consequently, there is a tendency of having higher concentration of CO_2 in a very hot environment. The carbon monoxide (CO) concentration level in the Teaching Hospital was lower than in the Specialist. The level of natural ventilation in the Teaching Hospital with wider window openings is higher which reduces the concentration of CO in the building spaces. On the whole, both hospitals have their CO concentration level within the 3 months period below the maximum threshold limit of 9 ppm.

From the measurements of the IEQ parameters conditions in the selected hospital ward buildings, it is evident that the quality of the indoor environment differ substantially depending on the ward design configuration and orientation, and also the outdoor weather condition. The indoor environment in the hospital wards had different thermal and lighting conditions because of variations in orientations, window sizes and air inlet/outlet. Heat gains from radiation rays of sunlight vary with the building orientation, also affected the indoor thermal quality in each of the hospital ward buildings within the period of measurements. The Teaching Hospital wards whose orientation (NW-SE) allows the fenestration façade to fall within the sun path, maximised it for daylighting within the wards. However, the wards attract more heat through their wide windows as compared to the Specialist Hospital wards. This results into increased in indoor air temperature which is however, minimised through the influence of natural cross ventilation provided in the Teaching Hospital wards. The Specialist Hospital on the hand, with closed-plan design configuration are not provided with proper natural ventilation through passive means to allow for free air flow and exchange. This contributed to the higher level of temperatures recorded especially in April, which were above the acceptable limits as specified by BS 15251 [28]. This also contributed to the higher levels of Carbon dioxide (CO_2) and Carbon monoxide (CO) concentration which affected the indoor air quality (IAQ). This result is in line with a study carried out by Altomonte et al. [48], which revealed that occupants in open-plan spatial layout buildings have a significantly higher level of satisfaction with their IEQ.

5. Conclusion

Indoor environmental quality (IEQ) problems in buildings are a result of certain decisions made during the building design processes and construction. As much as some of these problems can be solved through corrective measures such as retrofitting, it is essential to prevent and correct such deficiency at the building design stage, which is more economical and cost-effective. The outcome of this study would therefore serve as feedback to architects in the deign process leading to improvement in sustainable hospital ward design.

The essence of a hospital as a healthcare facility is to provide an environment that promotes healing rather than the one that hinders it. The design and maintenance of hospital ward buildings should be such that the main building occupants (patients) would feel homely throughout their period of stay within the hospital environment. The objective measurement of IEQ in the selected hospital wards can be said to be substantially different according to the ward buildings orientation and design configurations. The hospital wards having Northwest-Southeast orientation and open-plan configuration had a better IEQ as seen from the results. The Teaching Hospital wards as expected has better IEQ since its design, configuration and orientation is more environmentally friendly as compared to the Specialist Hospital. The internal wards layout in the Specialist Hospital may have contributed to ventilation problem, which affected the thermal and air quality considerably. Hospital ward building orientation and design can therefore be harness towards reducing heat gain and providing natural ventilation, which can also reduce energy demand for cooling especially in tropical Nigeria. In conclusion, the design of hospital wards for improved IEQ conditions should be such that proper attention is given to the orientation, floor plan configuration and window design for natural ventilation and lighting.

Abbreviations and symbols

ASHRAE	American Society of Heating, Refrigeration, and Air-conditioning Engineers
BS	British Standards
BREEAM	British Research Establishment Environmental Assessment Method
CO	carbon monoxide
CO_2	carbon dioxide
CIBSE	The Chartered Institution of Building Services Engineers
dBA	decibel
GBCA	Green Building Corporation of Australia
IAQ	indoor air quality
IEQ	indoor environmental quality
LEED	leadership in energy and environmental design
Lux	illuminance
Min	minimum
Max	maximum
NE-SW	Northeast-Southwest
NRC	noise reduction coefficient
NW-SE	Northwest-Southeast
PPM	parts per million
SD	standard deviation
WHO	World Health Organisation

WWR	window wall ratio
°C	degree Celsius
%	percentage

Author details

Pontip Stephen Nimlyat*, John James Anumah, Michael Chijioke Odoala and Gideon Koyan Benjamin
Department of Architecture, Faculty of Environmental Sciences, University of Jos, Nigeria

*Address all correspondence to: pontipn@unijos.edu.ng

References

[1] Garnys V. Indoor environment quality, design, and the value of facility ecology, Environment Design Guide (Tec 22). 2007:1-6. Retrieved from http://www.yourbuilding.org/library/1_TEC22.pdf

[2] Bonda P, Sosnowchik K. Sustainable Commercial Interiors. Hoboken, New Jersey: John Wiley & Sons; 2007

[3] Sarbu I, Sebarchievici C. Aspects of indoor environmental quality assessment in buildings. Energy and Buildings. 2013;**60**:410-419

[4] Taylor P, Sakhare VV, Ralegaonkar RV. Indoor environmental quality: Review of parameters and assessment models. Architectural Science Review. 2014;(March 2015):37-41

[5] Nimlyat PS, Kandar MZ. Appraisal of indoor environmental quality (IEQ) in healthcare facilities: A literature review. Sustainable Cities and Society. 2015;**17**:61-68

[6] Asadi I, Mahyuddin N, Shafigh P. A review on indoor environmental quality (IEQ) and energy consumption in building based on occupant behavior. Facilities. 2017;**66**(11/12):684-695

[7] Piasecki M, Kostyrko K. Indoor environmental quality assessment: Part 1: Choice of the indoor environmental quality sub-component models. Journal of Building Physics. 2017;**4**(3):264-289

[8] Mui KW, Chan WT. A new indoor environmental quality equation for air-conditioned buildings. Architectural Science Review. 2005;**48**(1):41-46

[9] Qi M, Li X, Zhu E, Shi Y. Evaluation of perceived indoor environmental quality of five-star hotels in China: An application of online review analysis. Building and Environment. 2017;**111**(July):1-9

[10] Al-Rajhi SMM, Ramaswamy M, Al-Jahwari F. IAQ in hospitals—Better health through indoor air quality awareness. Texas A&M University: Energy Systems Laboratory; 2010. Available Electronically from http://hdl.handle.net/1969.1/94139

[11] Zimring C, DuBose J. Healthy healthcare settings. In: Dannenberg AL, Frumkin H, Jackson RJD, editors. Making Healthy Places: Designing and Building for Health, Wellbeing and Sustainability. Washinton: Island Press; 2011. pp. 203-215

[12] Guenther R, Vittori G, Atwood C. Values-driven design and construction: Enriching community benefits through green hospitals. In: Designing the 21st Century Hospital Environmental Leadership for Healthier Patients and Facilities. Concord CA: The Center for Health Design, Health Care without Harm, and Robert Wood Johnson Foundation; 2006. http://community-wealth.org

[13] Langdon D. The cost and benefit of achieving green buildings. In: Info Property and Construction Data: Innovative Thinking; 2007. Retrieved August 30, 2015, from www.davislangdon.com

[14] Phelps A, Horman M, Barr M, Brower J, Riley D, Vanegas J, et al. Bridging the physics of building with the physiology of healthcare: Green healthcare facilities. Journal of Green Building. 2006;**1**:164-176

[15] Kim S, Osmond P. Analyzing green building rating tools for healthcare buildings from the building user's perspective. Indoor and Built Environment. 2014;**23**(5):757-766

[16] Alalouch CR. Hospital ward design: Implications for space and privacy

[PhD thesis]. Edinburgh: Heriot-Watt University; 2009

[17] Cox A, Groves P. Hospital and Healthcare Facilities. A Design and Development Guide. London: Oxford: Butterwoth-Heinaman; 1990

[18] Yau YH, Chandrasegaran D, Badarudin A. The ventilation of multiple-bed hospital wards in the tropics: A review. Building and Environment. 2011;**46**(5):1125-1132

[19] Castro MF, Mateus R, Bragança L. The importance of the hospital buildings to the sustainability of the built environment. 2012. Retrieved November 4, 2014, from http://repositorium.sdum.uminho.pt/handle/1822/21442

[20] Kothari C. Research Methodology: Methods and Techniques. 2nd ed. New Delhi: New Age International Publishers; 2004

[21] National Population Commission. Population distribution by sex, state, LGA & senatorial district. 2010. Retrieved from http://www.population.gov.ng/images/Vol03-Table-DSx-LGAPop-by-SDistrict-PDF.pdf

[22] NiMet. Current Climate Review Bulletin. Abuja Nigeria: Nigerian Meteorological Agency (NiMet); 2015

[23] Agwo YF, Wannang NN. Doctor-pharmacist collaborative role in patient management: Perception of patients, doctors and pharmacists. West African Journal of Pharmacy. 2014;**25**(1):55-67

[24] James WP, Tatton-Brown W. Hospitals: Design and Development. London: Architectural Press; 1986

[25] Zain-Ahmed A, Sopian K, Othman M, Sayigh A, Surendran P. Daylighting as a passive solar design strategy in tropical buildings: A case study of Malaysia. Energy Conversion and Management. 2002;**43**(13):1725-1736

[26] Ninomura BP, Ashrae M, Rousseau C, Ninomura P, Bartley J. Design and construction of hospital and health care facilities. ASHRAE Journal. 2006;**48**(June):33-37

[27] ASHRAE. ANSI/ASHRAE addendum F to ANSI/ASHRAE standard 62.1-2004. In: American Society of Heating, Refrigerating and Air-Conditioning Engineers. USA: ASHRAE Atlanta; 2006. pp. 1-4

[28] British Standards Institution. BS EN 15251. London: British Standard Institution (BSI) Publication; 2007. pp. 1-52

[29] Nicol JF, Humphreys MA. Adaptive thermal comfort and sustainable thermal standards for buildings. Energy and Buildings. 2002;**34**(6):563-572

[30] ASHRAE. ANSI/ASHRAE Standard 55. Atlanta, USA: American Society of Heating, Refrigeration, and Air-conditioning Engineers, Inc.; 2004

[31] World Health Organization (WHO). Guidelines for Community Noise. No. April. Geneva, Geneva: World Health Organization; 1999

[32] ASHRAE. ASHRAE Applications Handbook. Atlanta, USA: American Society of Heating, Refrigeration, and Air-conditioning Engineers, Inc.; 2007

[33] National Building Code (NBC). Lighting and Ventilation. India: Part 8; National Building Code of India; 2016. http://bis.org.in/sf/nbc.htm

[34] CIBSE. Lighting Guide: Hospitals and Healthcare Buildings. London: United Kingdom; 1989

[35] ASHRAE. ANSI/ASHRAE 62. 1-2010. Atlanta, USA: American Society of Heating, Refrigeration, and Air-conditioning Engineers, Inc. USA: ASHRAE Atlanta; 2010

[36] British Standards Institution. BS EN ISO 28802. London United Kingdom

British: Standard Institution (BSI) Publication; 2012

[37] R. de Dear and G. S. Brager, "The adaptive model of thermal comfort and energy conservation in the built environment," International Journal of Biometeorology, vol. 45, pp. 100-108, 2001.

[38] EPA. Ambient air quality standards (NAAQS). In: U.S. Environmental Protection Agency (EPA). USA: California Air Resource Board; 2011. www.arb.ca.gov/research/aaqs/aaqs/naaqs/

[39] FGI/ASHE. Sound and vibration design guidelines for health care facilities. Public Draft 2.0. 2010. Retrieved from http://www.speechprivacy.org

[40] ASHRAE. ASHRAE Guideline 10P. Atlanta, USA: American Society of Heating, Refrigeration, and Air-conditioning Engineers, Inc.; 2010

[41] Malaysian Standard. MS1525. Malaysia, Malaysia: Department of Standards; 2001

[42] Reinten J, Braat-eggen PE, Hornikx M, Kort HSM. The indoor sound environment and human task performance: A literature review on the role of room acoustics. Building and Environment. 2017;**123**:315-332

[43] Kibert CJ. Sustainable Construction: Green Building Design and Delivery. 3rd ed. New York: John Wiley & Sons, Inc.; 2012

[44] Kim G, Kim JT. Healthy-daylighting design for the living environment in apartments in Korea. Building and Environment. 2010;**45**:287-294

[45] CIBSE—LG2. Lighting for Hospitals and Healthcare Buildings. England: The Chartered Institution of Building Services Engineers (CIBSE); 2008. www.cibse.org/knowledge-item/details

[46] BS EN 12464-1. Light and Lighting - Lighting of Workplaces. British Standard Institution (BSI) Standard Publication; 2011

[47] NEA-SS554. National Environmental Agency: Code of Practice for Indoor Air Quality for Air-Conditioned Buildings. Singapore: Spring Singapore; 2009

[48] Altomonte S, Saadouni S, Kent MG, Schiavon S, Altomonte S, Saadouni S, et al. Satisfaction with indoor environmental quality in BREEAM and non-BREEAM certified office buildings office buildings. Architectural Science Review. 2017;**60**(4):1-13

Chapter 6

Sound Quality inside Mosques: A Case Study on the Impact of Mihrab Geometry

Hany Hossameldien and Abdulrahman Abdullah Alshawan

Abstract

This chapter presents a detailed analysis to identify the best Mihrab (prayer niche of Imam) geometry with respect to acoustic performance of mosques. Mihrab geometry has an impact on daily prayer recitation and orders, as the Imam (prayer leader) faces this semicircular shape. Sound pressure level (SPL) has been simulated to compare different well-known designs and geometries of Mihrab by ODEON. A typical mosque in an educational campus was considered for the study, and it is found that Safavid Mihrab geometry has the best performance followed by Chinese, Mughal, and Tulunid Mihrab geometries, while the worst performance is found in Almoravid Mihrab geometry. This study obviously guides the mosque designers to choose the appropriate Mihrab geometry with regard to the acoustic performance.

Keywords: mosque acoustics, prayer niche, Mihrab geometries

1. Introduction

The term masjid (mosque) refers to the "place for prostration," which is used by Muslims as houses of worship. Muslims have to execute five prayers daily, which are supposed to be performed congregationally in masjids. Masjids are exclusively essential structures in every Muslim community. They normally have a certain size and location in relation to the public. In general, they could be categorized as large national masjids, major landmark buildings, community focal point, and small local neighborhood masjids. Although their uses are clearly varied, they have several consistent characteristics. Mihrab (prayer niche) is the prayer place in Arabic, and it is the architectural feature of front wall (to which the Imam faces during the daily prayer) of any masjid.

Few studies on mosque acoustics have been reported by Topaktaş [1]. In general, the literature on mosque acoustics could be categorized as follows [2]:

A. Researches on academic studies that concentrated on the analysis of existing mosques

B. Studies on acoustic renovation and modification

C. Studies on real or virtual mosques to develop acoustic design criteria

The first category (A) contains assessment researches including single mosque cases, comparisons of mosque to another mosque, or comparisons of mosques to churches or chapels. The second category (B) contains recommended architectural modifications on floor plan geometry and materials in contrast to common utilized internal finishes or shapes. Rare cases applied solely electronic sound reinforcement systems with no modifications on interior design, while both approaches were used in some cases. The third category (C) aims either to propose particular architectural parameters/features that are effective on the acoustics of the mosque typology or to specify acoustic parameter limits specifically to be applied for mosque typology [2].

1.1 Studies on analysis of existing mosques

These researches include assessment of single mosque or church cases, comparisons between mosques, and comparisons of mosques to churches or chapels. António and Carina [3] studied the acoustic performance of central mosque of Lisbon, Portugal. They measured the acoustic characteristics such as reverberation time (RT), rapid speech transmission index (RSTI), and background noise (BN). They did measurements for unoccupied situation in male and female prayer halls. The outcomes were analyzed and compared to another studies done for Catholic churches and mosques within volume average; in general, the average RT was 500–1 kHz but was a little higher when compared to the value recommended by the authors. El-khateeb and Ismail [4] studied the speech intelligibility in Sultan Hassan Mosque and Madrasa situated in Cairo, Egypt, by field measurements and ODEON simulation. The parameters were RT and Speech Transmission Index (STI) for occupied and unoccupied cases. They concluded that Sultan Hassan Mosque and Madrasa had high RT and echo at some examined points; however, it did not impact worshippers' understanding of Imam either in Friday speech or daily prayers.

Zühre and Yilmazer [5] investigated the acoustic characteristics of Kocatepe mosque in Ankara, Turkey, and compared them with masjid in the ancient Othman period. Kocatepe had a long RT in low frequencies due to central dome which was the aim of the study. An analysis and computer simulation by ODEON 6.05v were carried out to identify the acoustic features, including the parameters such as RT, early decay time (EDT), clarity (C80), sound definition (D50), lateral fraction (LF), STI, and strength (G). They tested three scenarios: the empty mosque, prayer mode when mosque was one-third full, and fully occupied, for daily and Friday speech. The acoustic performance of Kocatepe mosque was below average when empty but was acceptable when entirely occupied. António and Cândido [6] did an acoustic comparison of Catholic churches and mosques to clarify the main similarities and differences based on architectural and acoustic features. They studied variabilities between the following parameters for unoccupied spaces: RT, C50/C80, and STI/RSTI. Also, they considered the architectural information related to each building such as volume, length, area, height, and width. From measurements on churches (41 buildings in Portugal) and mosques (21 buildings in Saudi Arabia), they concluded that mosques in general had an overall better acoustic performance due to floor surface absorption value. A similar study was also reported from Istanbul, Turkey [7], which compared four mosques and three churches in Turkey. António and José [8] investigated the acoustics of Mekor Haim Synagogue (Jewish worship place), Portugal. The aim of this study was to compare the acoustic behavior of the Synagogue with Catholic churches in Portugal and mosques in Saudi Arabia with comparable volume. They suggested reducing RT only at dome in order to enhance the Synagogue acoustics. David and Paulo [9] evaluated the acoustic performance of a contemporary church in Curitiba, Brazil, to study its compliance with NBR 12179 Brazilian National Standard, ISO 3382-1 international standard and IEC 60268-16 standard. They measured RT and D50 in accordance with ISO 3382 and ISO 3382-1 and calculated STI by ODEON software. It was found that the overall performance of the church exceeded the recommended values of some standards and was satisfactory for some parameters in a specific standard. An acoustic comparison of modern and ancient mosques has been done by Zerhan and Sevda [10] for a modern mosque and an ancient mosque.

1.2 Studies on architectural features and recommendation of floor plan geometries and materials

Few studies applied solely electronic sound reinforcement systems (SRS) with no modifications on interior design, while in some, both approaches are used. Abdou [11] made a wide analysis of the most common mosque floor plan geometries to measure the effect of the floor plan geometry on acoustic performance, particularly on the spatial distribution patterns of speech intelligibility without SRS. A simulation has been done of sound fields of five common forms of Muslim worship activities and level of occupancy. The study concluded that the square floor plan was the best in terms of acoustics. Another study [12] focused on Mihrab design and its basic acoustic characteristics of traditional vernacular mosques in Malaysia. The aim of this study was to review the acoustic performance of the investigated mosques and also to evaluate the acoustic performance of Mihrabs. The researchers surveyed 37-year-old mosques built within the period 1728–1830 in Malaysia; all these mosques had either square or rectangular floor plan geometry. The Mihrabs of the investigated mosques had circular niche with flat ceiling to rectangular shape with slanting ceiling and semicircular concaved niched forms. They utilized a PC-based acoustics measuring system and analyzer, and data from previous five case studies were analyzed and compared. They concluded that Mosque Mihrabs offered a good feature to confirm the trend of reasonable acoustic performance with a maximum variance of 4.0 dB. Utami [13] studied about domes coupled to rooms in mosques to identify the impact of centralized ceiling domes on acoustic performance of mosque buildings. A computer model was developed to compare the outcomes derived from analytical, numerical, and experimental (scale modeling) methods. Moreover, statistical techniques such as ANOVA and t-tests were utilized to compare the experimental results. The conclusion was based on comparisons and on realization listening tests in order to discover mosque components that produced the major acoustic impact. The analysis could establish criteria for better mosque acoustic performance with domed ceiling. Kayili [14] examined the applied acoustic systems throughout the history in Othman period; the study elaborated domes and cavity resonator technology made of bronze as well. The researcher found a variety of plaster types on internal surfaces of the investigated mosque (Selimiye mosque in Istanbul, Turkey).

A study about the influence of SRS on acoustic performance in churches [15] analyzed the sound field and its influence to speech intelligibility and clarity of music and recitation. The acoustic parameters such as RT, EDT, D50, C80, and STI were measured with the impulse response technique and compared the outcome with and without SRS. It was shown that SRS improved D50 and C80 for sound receivers. Also, for EDT the reverberance sensation decreased by distance reduction between sound receiver and source. The study concluded that SRS could provide slight enhancement in speech and music/recitation perception; however, it did not solve the issue originated by poor acoustic design. A similar work [16] was also reported on acoustics in worship spaces particularly mosques containing existing or newer computer-supported SRS. The main goal of this study was the development and optimization of control algorithms of such systems using digital signal processing (DSP) controlled electroacoustic devices and computer-based systems to reach required radiation properties. In terms of floor plan geometry, Eldien and Al Qahtani [17] studied the most common two geometries of mosques, which are rectangular and square. They excluded the dome, worshippers, and sound reinforcement system and used the same finishing materials for both shapes for proper and fair simulation. They measured reverberation time (RT), early decay time (EDT), and sound transmission index (STI). This study found that the square floor plan has better overall sound qualities.

1.3 Studies on acoustic parameters and design guidelines

In this section, previous studies on particular architectural parameters/features that affect the acoustics of the mosque/worship spaces' typology and/or specifying acoustic parameter limits specifically to be applied for mosque typology are included. Abdou [18] studied the acoustic characteristics of existing Saudi Arabian mosques, by conducting field measurements (for parameters such as RT and C50) in 21 typical mosques which had diverse sizes and architectural features. The aim was to list down or specify their acoustic performance and to clarify air cooling system, ceiling fans, and sound systems' acoustic effect. BN was measured with and without air conditioning system operation, while STI was evaluated with and without SRS. It was deduced that the acoustic qualities of the investigated mosques deviated from optimum conditions when it was empty, but the acoustic performance improved in the occupied condition. Similar study on measuring STI with and without SRS was also reported by Cunha [15] for a church.

Diocese of Columbus [19] provided acoustic recommendations for the construction and renovation of churches and chapels. The study clarified the most important factors for acoustic design as listed below:

1. Basic requirements for good acoustics in churches

2. Elements of good natural acoustics for worship

3. Physical provisions for sound source isolation

4. Mechanical system noise control

5. Sound reinforcement system acoustics and the design/building process

6. An acoustic checklist for a typical church building process

In addition, it suggested guidelines for an appropriate natural acoustics in the architectural and acoustic design of churches and chapels, which included to provide RT of at least 2–3 s and to minimize the amount of sound-absorbing materials. In all cases, sound-absorbing materials should not be situated nearby the important sources of sound: the assembly, the music ministry, and the presiders and readers. Since all of these sound sources are at the floor level, floors cannot be carpeted in churches, and pews cannot be covered with upholstery or cushions. Additional suggestions included providing sufficient room volume to allow the natural development and support of sound. A volume of 300–400 cubic feet per seat was recommended for churches with seating capacities up to about 800 seats. Larger churches might require greater volume, but smaller churches should not fall substantially below this range. In providing sufficient room volume for acoustics, height is a more important factor than floor area. It was also suggested to provide properly oriented, hard-surfaced materials around sound sources. All surfaces (including floors, walls, and ceilings) near and around presiders, cantors, readers, musicians, and the assembly must have hard surfaces. The study concluded that the acoustic effort includes the four essential facets of church acoustics: (1) natural/architectural acoustics, (2) sound isolation, (3) mechanical system noise and vibration control, and (4) sound reinforcement system design and specification. The following list of acoustic checkpoints was provided for acoustic consultants:

1. Predesign and programming phases

2. Schematic design phase

3. Design development phase

4. Construction document phase

5. Construction phase

6. Final construction evaluation

Each step of the above checklist has its own requirement that helps any designer to generally manage the acoustic requirements from project designing phase. Besides, the study gave general instructions that could be used for any building without parametric specification or limitations. Francesco et al. [20] provided guidelines for acoustic measurements in churches, with the motive of preserving the architectural features of this group of cultural heritage buildings. A team of three Italian universities was formed to provide technical and operative supports to perform measurements inside different churches. They collected detailed data of acoustic features of most important cultural heritages in order to improve the knowledge of sound propagation, preserve the architectural aspects in case of renovation, and determine the best approaches to improve the acoustic performance in existing buildings. A set of guidelines were proposed to simplify and normalize the choice of source and receiver locations and to suggest suitable hardware for acoustic measurement in churches.

Ismail [21] highlighted that designers do not pay enough attention to the acoustic performance in prayer halls due to projects' time limitation and insufficient basic guidelines for better acoustic performance during the early design stage. He investigated the acoustic performance of contemporary mosques by using computer model based on ray tracing theory. The study considered three most common mosque design topologies, which had different size, shape, and internal surface materials. Diverse acoustic treatments were studied to the geometric nature. The study provided design recommendations and guidelines that could help architects in conceptual design. A case study by Zühre [2] in Dogramacızade Ali Pasa mosque (Ankara, Turkey) focused on the impact of design decisions on acoustic comfort parameters. The selected mosque had a unique design, which was intensively studied at all design phases. Simulations were done by ODEON v10.13, and the outcomes were validated by site measurement in the studied space. The acoustic parameters assessed were RT C50, C80, STI, and sound pressure level (SPL).

1.4 Summary of literature review

In the first category, the studies focused on existing unoccupied mosques such as Lisbon mosque in Portugal and Sultan Hassan Mosque in Egypt. Moreover, these studies compared the acoustic performance between mosques and churches with comparable characteristics and volumes. Besides, some studies were on existing churches to evaluate the acoustic performance. In the second category, which focused on renovation and modification of worship places, the most common mosque floor plan geometries were studied and compared. In addition, these studies addressed the acoustic impact of the architectural elements of mosques or churches, such as Mihrabs, domes, and internal finishing, and the influence of SRS on acoustic performance in churches. The studies in the third category were on virtual or real mosques to develop acoustic design criteria. In general, the floor plan geometry of worship spaces was investigated. In addition, many researches were reported about different shapes and styles of domes and other architectural features in mosques and churches, which have a major effect in room acoustics. Some studies were about acoustic performance of existing Mihrab. However, detailed research on the acoustic performance of well-known Mihrab geometries is still lacking, which is the focus of this chapter.

2. Methodology

The simulated scenario is the daily prayer when the mosque is empty, by following ISO acoustic standards and procedures (building acoustic measurements standards ISO 140 and ISO3382). We have developed a three-dimensional design of a typical mosque inside Imam Abdulrahman Bin Faisal University (IAU) campus of Saudi Arabia and imported into the simulation tool (ODEON). The simulations were performed for 10 Mihrab geometries whose data were collected from previous studies. Mihrab's height and width were fixed. The SPL values were analyzed in order to select the best Mihrab geometry in terms of acoustic performance.

2.1 Mihrab development throughout Islamic history

Islamic architecture developed throughout Islamic empire expansion, the massive Islamic land from Eastern Asia toward Africa and some parts of Europe, has influenced mosque component architecture such as Mihrab, Minarat, and Quba [22]. **Table 1** shows each Islamic period that contributed on mosque architecture and development. Each Islamic period in **Table 1** has a masterpiece mosque

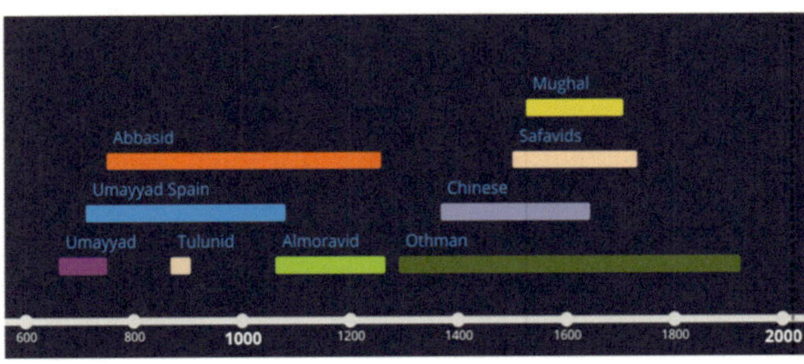

MOSQUE NAME	PERIOD
Umayyad	661 - 750
Umayyad Spain	711 - 1081
Abbasid	750 - 1258
Tulunid	868 - 905
Almoravid	1062 - 1265
Othman	1290 - 1922
Safavids	1501 - 1732
Mughal	1526 - 1707
Chinese	1368 - 1644

Table 1.
Mihrab development throughout Islamic history (historical timeline).

Period	Nation	Masjid	Mihrab Floor Plan	Mihrab Elevation
661-750	Umayyad	Great Masjid of Damascus		
711-1031	Umayyad Spain	Great Masjid of Cordoba	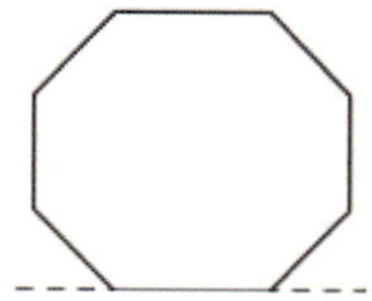	
750-1258	Abbasid	Abu Dulaf Masjid		
868-905	Tulunid	Masjid of Ibnu Tulun		
1062-1269	Almoravid	Great Masjid of Tlemchen	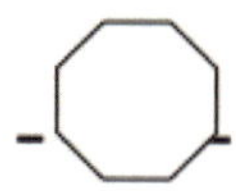	
1290-1922	Ottoman	Üç Serefeli Masjid		
1501-1732	Safavids	I-Shah Masjid		
1526-1707	Mughal	Moti Masjid	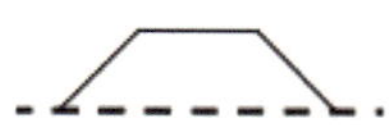	
1368-1644	Chinese Dynasty	Great Masjid of Xi'an		
Recent	Saudi Arabia	IAU Masjid (Base Case)		

Table 2.
Development of Mihrab shape throughout Islamic history.

(landmark) of the nation describing their architecture and culture. **Table 2** presents each period's famous Mihrab geometry and related information.

2.2 Modeling configurations

The various Mihrab geometries used for the computer simulation are summarized in **Table 2**. The size of the mosque corresponds to a community mosque with prayer hall area of 28 × 28 m and ceiling height of 4.95 m, as illustrated in **Figure 1**.

One worship scenario is examined in the simulation. The congregation (worshippers) performs the prayer behind the Imam who recites in a standing position facing the Qibla niche using his raised voice. It is natural that persons delivering speech without the aid of electroacoustic sound system tend to raise their voice. The background noise in the mosque is assumed to reach a noise criterion (NC) rating of NC30 (religion spaces). The worshippers are assumed to be listening to the Imam

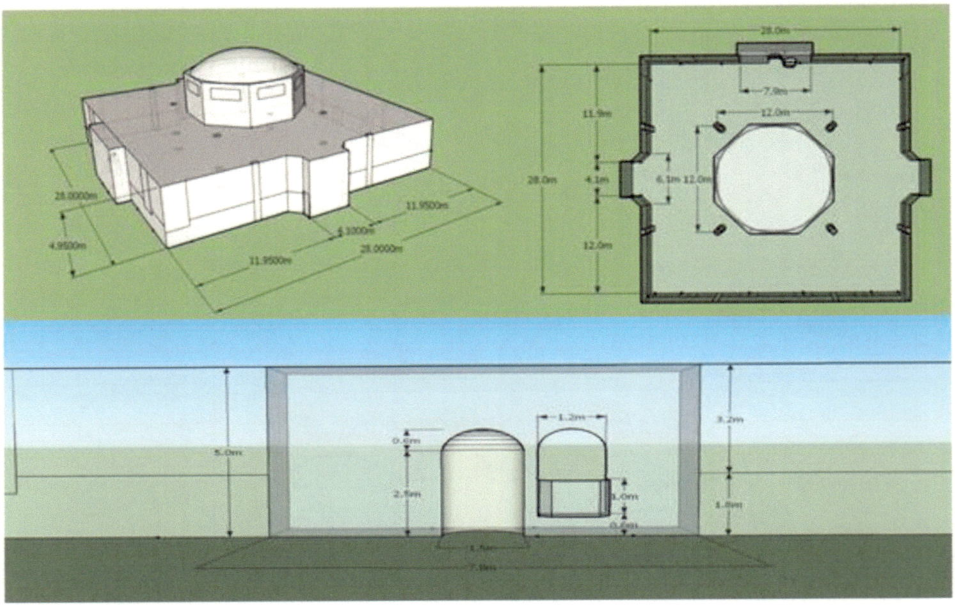

Figure 1.
IAU campus mosque dimensions.

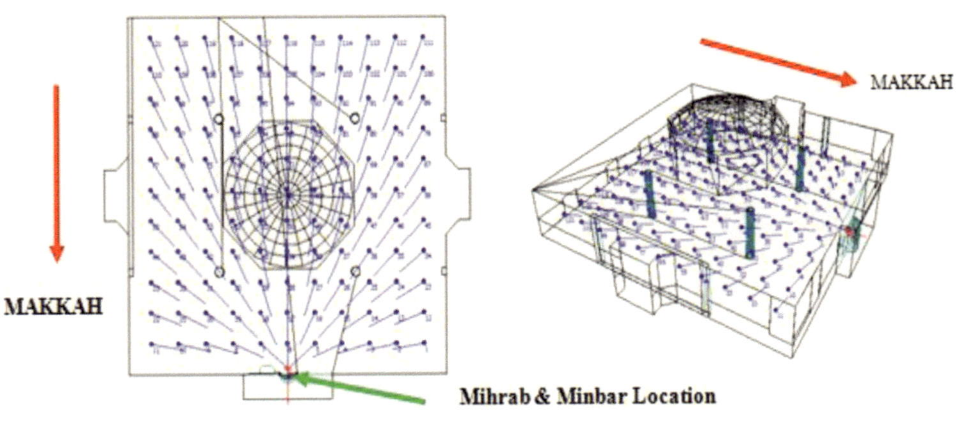

Figure 2.
Sound source and receiver points.

in standing position as is usually the case during the daily prayers. Their ear height was taken to be 1.65 m from the floor. Measurements and simulation were done when the mosque was assumed empty. **Figure 2** demonstrates the positions of sources and receives points for all configurations. These parameters were simulated in 121 different point positions as indicated in **Figure 2**. These points were measured for 10 Mihrab geometries including base case (**Table 2**). The distribution was on a 2.4 m grid. Each receiver point was 1.65 m high.

3. Results and discussion

SPL is the result of the pressure variations in the air achieved by the sound waves. The lowest sound pressure that can be heard by humans is called the hearing threshold, and the highest is known as the pain threshold. The human voice's average (speech) SPL is 50–70 dB. We investigated the average values for the selected Mihrab geometries on the overall floor plan and prayer row wise. Moreover, the analysis for Mihrab geometries was made on 2.4 m grid scale with 121 receiver's points distributed equally on the floor plan. When the SPL value is higher, it is an indication of higher acoustic performance.

Figure 3 presents the SPL contour plan for the Mihrab geometries studied. It presents the SPL performance for 10 Mihrab geometries. Therefore, each Mihrab geometry was investigated separately and, for each prayer row, with respect to the average SPL value. Finally, we summarized the SPL average values for all the prayer rows in order to compare them between the Mihrab geometries.

3.1 Base case Mihrab geometry SPL analysis (IAU)

Figure 4 shows the SPL contour values for the IAU (base case) Mihrab. We can observe that the SPL values decrease with the distance from the sound source. The

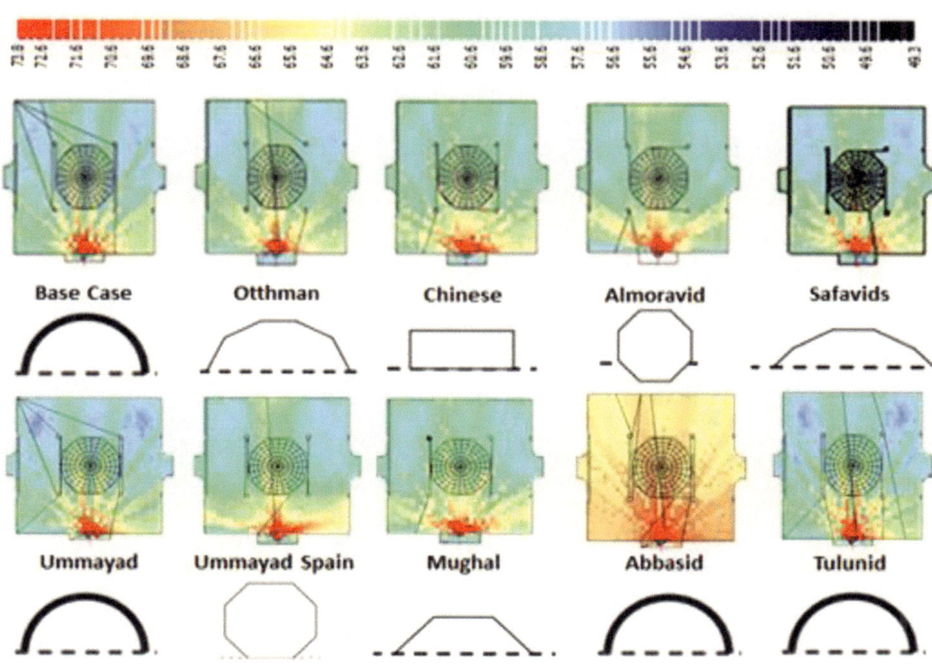

Figure 3.
SPL contour plans for all studied Mihrab geometries.

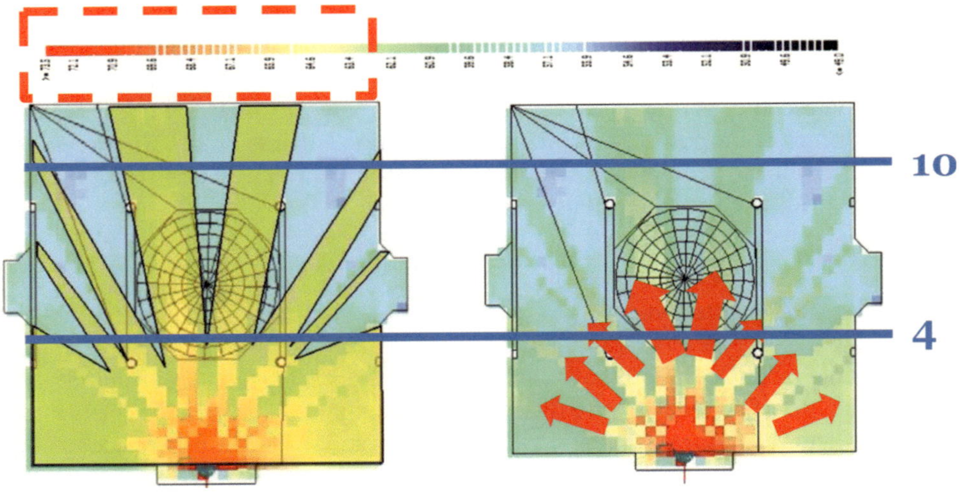

Figure 4.
SPL value contour plans for base case mosque Mihrab geometry at 1 kHz.

SPL values range from 73 to 65 dB at 1 kHz. In general, the maximum values are located at the areas near the Mihrab and at the two sides. In general, this case has a noticeable increase of SPL value at the most critical point, which is behind the sound source. In addition, the Minbar (pulpit for the preacher) impact is low, and 35% of mosque area has low dB. Moreover, the yellow-hatched area is about 510 m^2 which is equivalent to approximately 65% of the floor plan. This Mihrab geometry is more suitable for a rectangular floor plan, once the sound distribution is homogeneous all over the floor plan.

Figure 5 presents SPL average values for prayer rows of the base case Mosque Mihrab geometry. The SPL average values are varying from 70 to 65 dB. The first row has the highest SPL value due to its relative position to the source. The second and third prayer rows have the same average of SPL values. The fourth prayer row has comparatively lesser average value than the fifth prayer row. Moreover, there is a gradual decrease of SPL average values from the fifth to the tenth prayer row. There is a noticeable increase of SPL average value at the last portion of mosque floor plan from lower SPL average value from the tenth prayer row to the

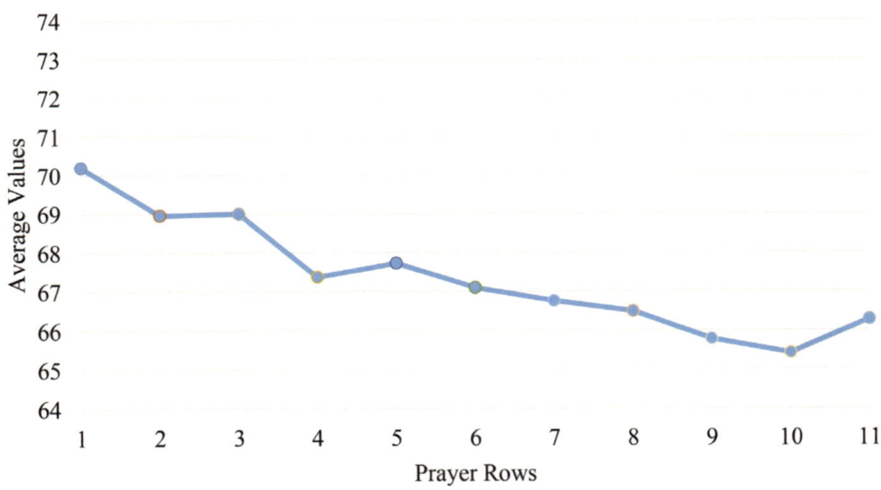

Figure 5.
SPL average values for base case Mihrab prayer rows.

eleventh prayer row. The average SPL value of this Mihrab geometry for 121 sound receiving points on the prayer rows is 0.65 dB.

3.2 Othman Mihrab geometry SPL analysis

Figure 6 illustrates SPL values of Othman Mihrab geometry. SPL values from 66 dB to 74 dB are near the sound source and start decreasing as we move away. In general, it is found that the maximum values are located at the middle of the row closest to the sound source. Moreover, the yellow-hatched area (60–74 dB) is about 470 m^2 which is equivalent to 59% of overall mosque floor plan. This Mihrab geometry is more suitable for rectangular floor plan once the direction of the sound (red arrows) is toward the plan's longer direction.

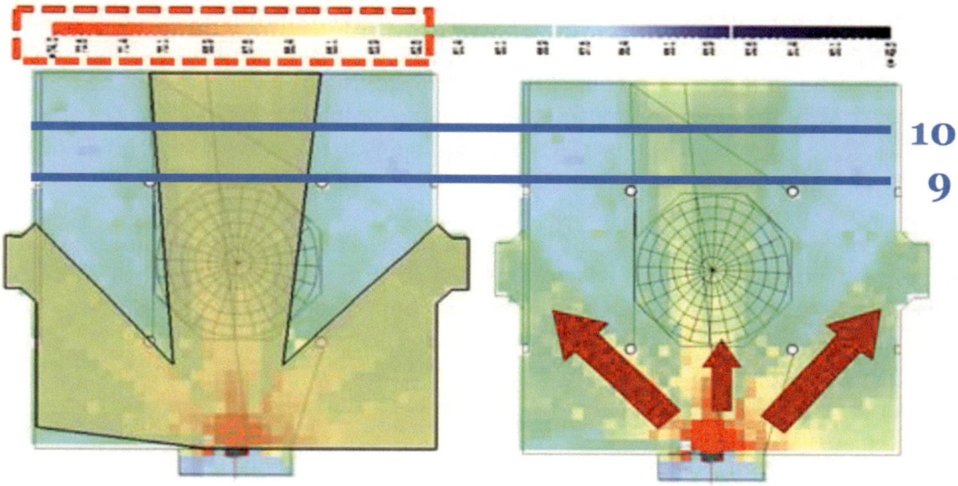

Figure 6.
SPL contour values for the Othman Mihrab at 1 kHz.

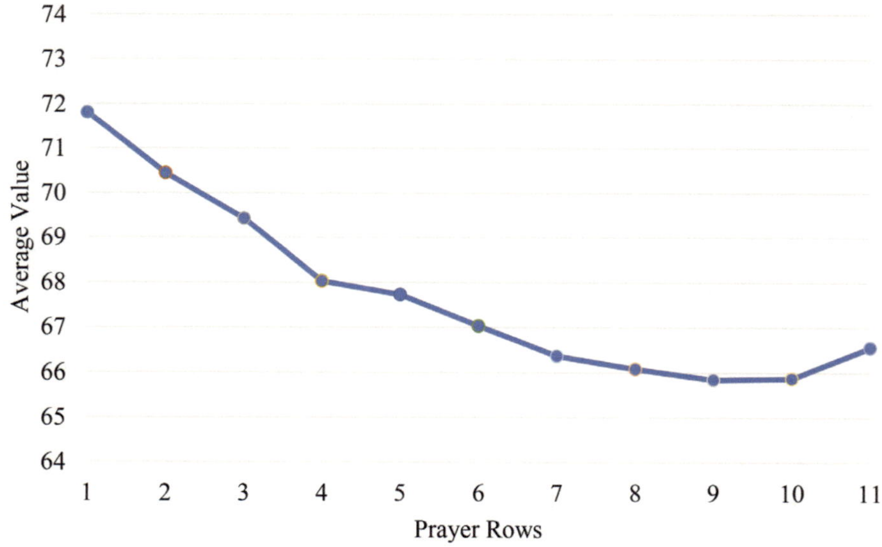

Figure 7.
SPL average values for Othman Mihrab prayer rows.

Figure 7 presents the SPL average values for prayer rows of Othman Mosque Mihrab geometry. The SPL average values are varying from 71 to 66 dB. The first prayer row has the highest SPL average value due to the sound source position. The SPL average values gradually decrease from the first prayer row to the lowest SPL average values at the ninth and tenth prayer rows. Moreover, the SPL average values are noticeably higher at the last portion which is found at the eleventh prayer row. The average SPL value of this Mihrab geometry for 121 sound receiving points on mosque prayer rows is 0.70 dB.

3.3 Chinese Mihrab geometry SPL analysis

Figure 8 shows the SPL values of Chinese Mihrab geometry. The SPL values are varying from 49 dB to 73 dB on the plan reference to SPL scale. **Figure 8** presents the scattered sound distribution of Chinese Mihrab geometry. The yellow-highlighted area (60 dB–73 dB) covers 62% of mosque floor plan. In general, this Mihrab geometry is a good sound distributor for the square and the rectangular mosques. The sound effect of this Mihrab is approximately the same as the Othman's Mihrab. **Figure 9** presents SPL average values for prayer rows of Chinese Mosque Mihrab geometry. The SPL average values are varying from 72 dB to 66 dB. The first prayer row has the highest SPL average value. The SPL average values are progressively decreasing from the first prayer row toward the tenth prayer row which has the lower SPL average value. The SPL average value steadily increases from the tenth to the last/eleventh prayer row. The average SPL value of this Mihrab geometry for 121 sound receiving points on mosque prayer rows is 0.70 dB.

3.4 Almoravid Mihrab geometry SPL analysis

Figure 10 demonstrates the SPL values of Almoravid Mihrab geometry. The SPL values are varying from 47 to 72 dB on the plan reference to SPL scale. Almoravid Mihrab has an octagonal shape. Almoravid Mihrab geometry has three

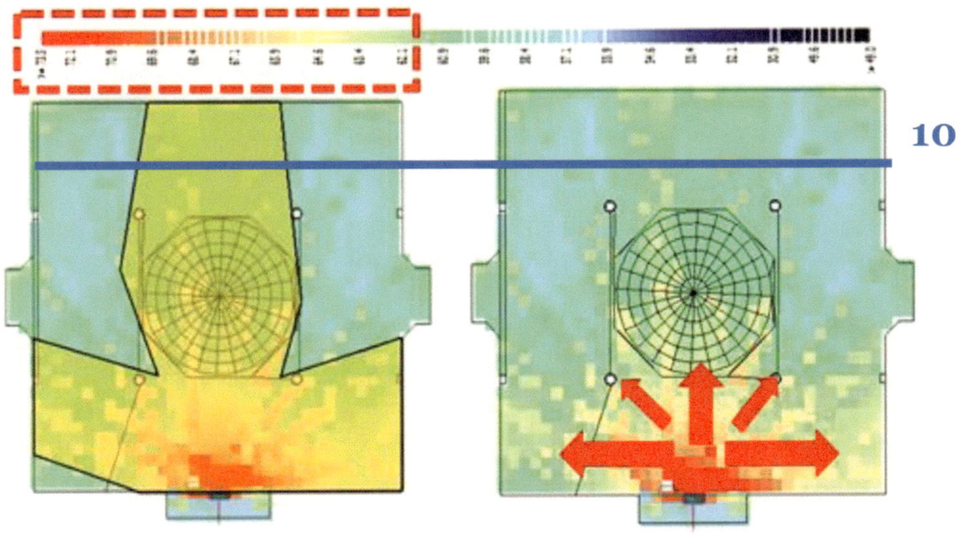

Figure 8.
SPL contour values for the Chinese Mihrab at 1 kHz.

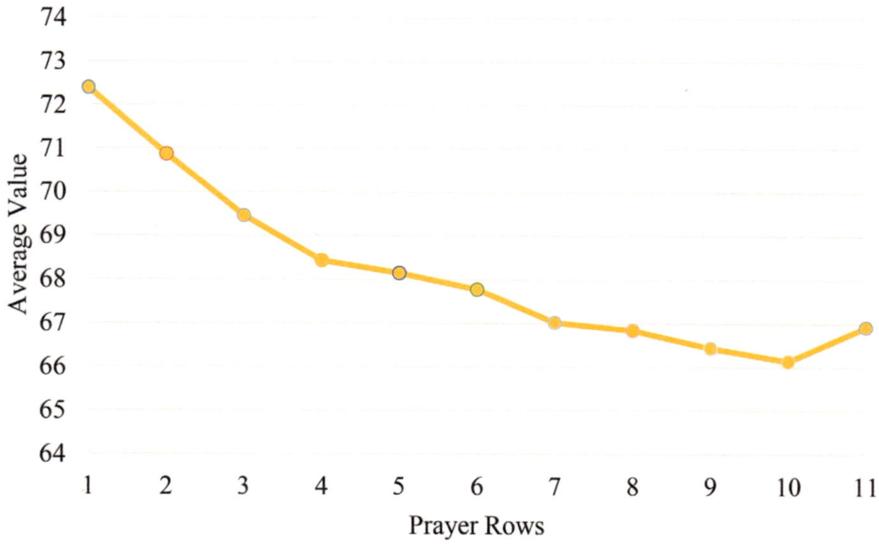

Figure 9.
SPL average values for Chinese Mihrab prayer rows.

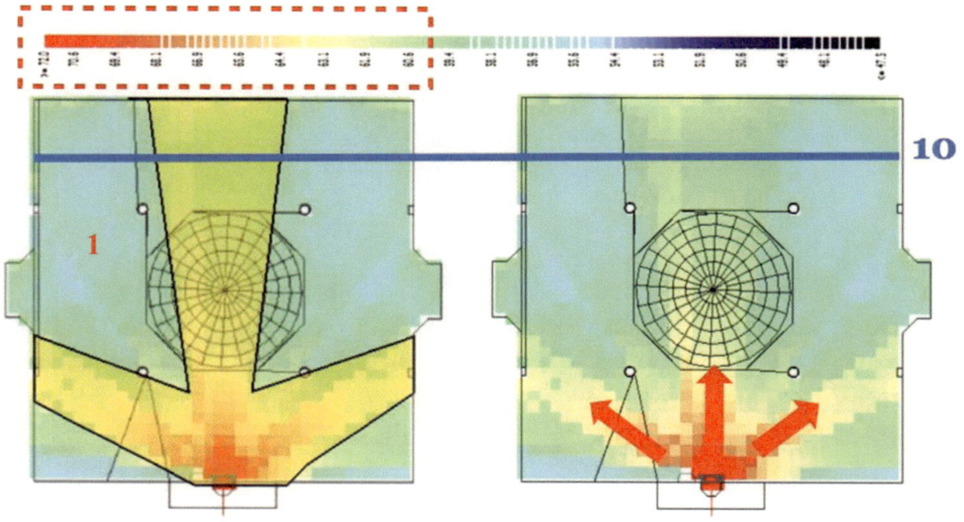

Figure 10.
SPL contour values for Almoravid Mihrab at 1 kHz.

main sound directions and very lower SPL dB throughout most of the floor plans. The sound effect of this Mihrab appears clearly at the center and the near sides of the mosque. The yellow-hatched (60–72 dB) area is 43% of plan surface. Almoravid Mihrab shape has generally low sound coverage on mosque receiving points.

Figure 11 illustrates SPL average values for prayer rows of Almoravid Mosque Mihrab geometry. The SPL average values are varying from 70.5 to 6.5 dB. The highest SPL average value is found at the first prayer row, and then it is gradually decreasing toward the tenth prayer row. Also, there is a little increase of SPL average value from the tenth to the eleventh prayer row. The average SPL value of this Mihrab geometry for 121 sound receiving points on mosque prayer rows is 0.65 dB.

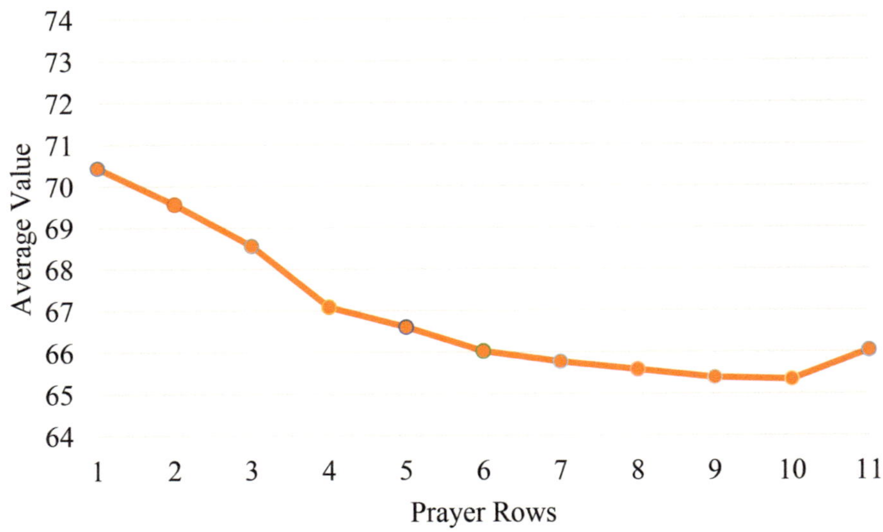

Figure 11.
SPL average values for Almoravid Mihrab prayer rows.

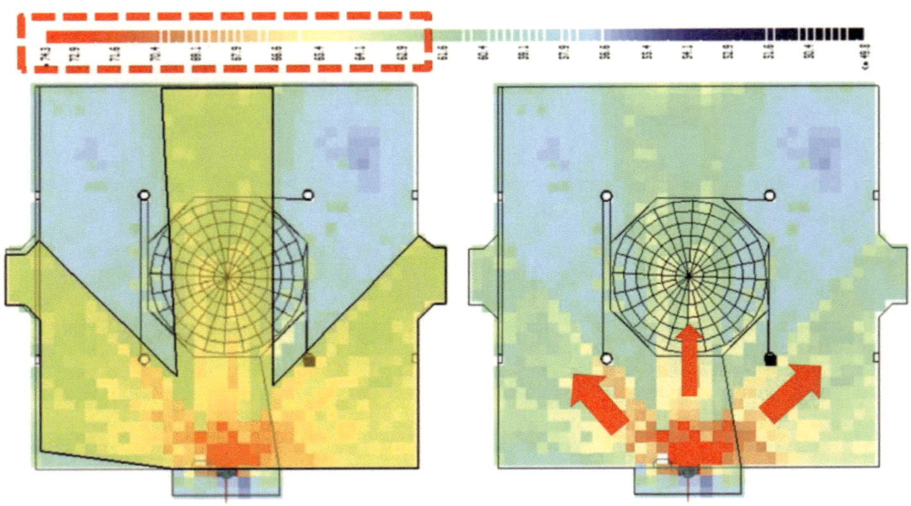

Figure 12.
SPL contour values for Safavid Mihrab at 1 kHz.

3.5 Safavid Mihrab geometry SPL analysis

Figure 12 shows the SPL values of Safavid Mihrab geometry. As shown in **Figure 12**, SPL values are varying from 49 to 77 dB. The highest SPL values are found in the area nearby Imam position (sound source) which is the first three prayer rows in general. The highlighted area covers 58% of the floor plan with good SPL values (60–74 dB). The maximum diffusion effect is located at the center and the first half area of the mosque. For that reason, Safavid Mihrab can have a good acoustic performance for the rectangular floor plan geometry mosques due to the main directions of sound reflection (red arrows).

Figure 13 presents SPL average values for prayer rows of Safavid Mosque Mihrab geometry. The average values vary from 37 to 66 dB. High SPL values are observed at the first prayer row and start decreasing till the tenth prayer row. The lowest average value is found at the tenth prayer row. Moreover, slight SPL average

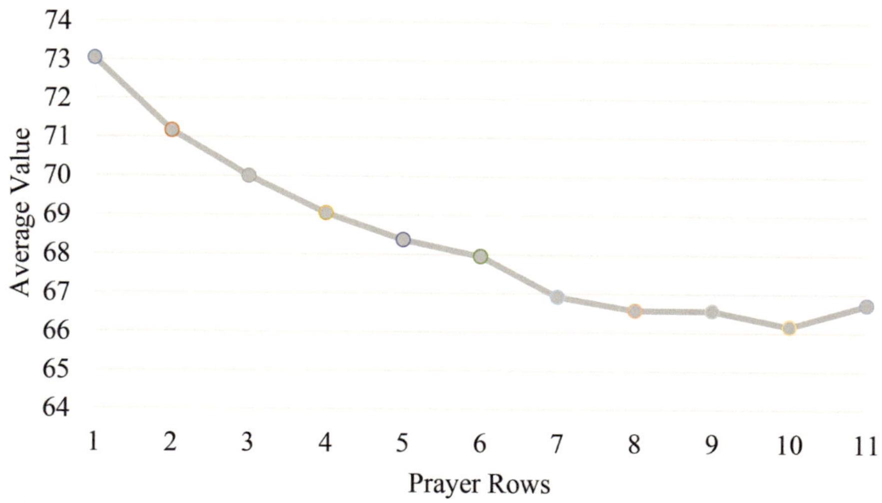

Figure 13.
SPL average values for Safavid Mihrab prayer rows.

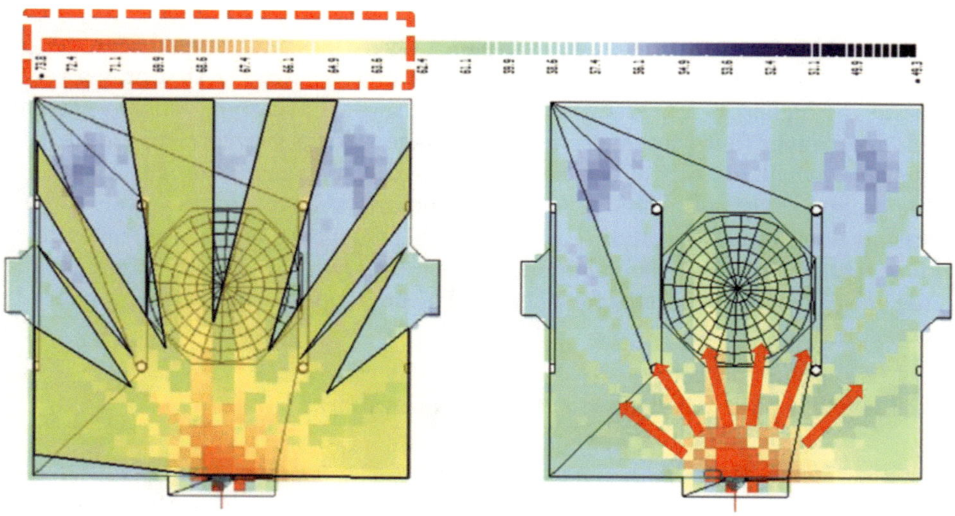

Figure 14.
SPL contour values for Umayyad Mihrab at 1 kHz.

value increment is noticed from the tenth to the last/eleventh and ninth prayer rows. The average SPL value of this Mihrab geometry for 121 sound receiving points on mosque prayer rows is 0.70 dB.

3.6 Umayyad Mihrab geometry SPL analysis

Figure 14 shows the SPL contours and SPL values for all proposed receiver points obtained by Umayyad Mihrab geometry. This Mihrab has a semicircle form as the IAU Mihrab. The main difference between the two Mihrabs is the architectural decorations and the top end of the Mihrab. For that, the results obtained by Umayyad Mihrab geometry are completely different from that obtained by the IAU Mihrab. Umayyad Mihrab diffuses the sound energy in a semicircle form, where the highest SPL values are found at the first three prayer rows nearby Imam position (sound source). The yellow-hatched area is covering 67% of floor plan area with

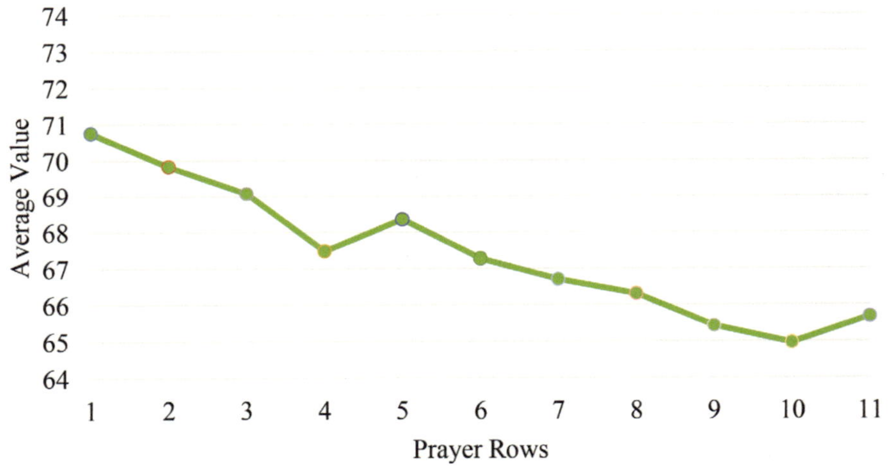

Figure 15.
SPL average values for Umayyad Mihrab prayer rows.

high SPL values ranging from 60 dB to 73 dB. Generally, this shape has scattered sound distribution on the floor geometry; thus, it is suitable for rectangular and square floor plans.

Figure 15 presents SPL average values for prayer rows of Umayyad Mosque Mihrab geometry. The average values vary from 70 to 65 dB. The first prayer row receives the highest SPL average value and then starts decreasing gradually toward the fourth prayer row. Moreover, the SPL average value is recovered at the fifth prayer row which has higher value than the fourth prayer row. The SPL average value progressively falls from the fifth to the tenth prayer row which receives the lowest average values. Another recovery on SPL average value is at the eleventh prayer row which has higher value than the tenth prayer row. The average SPL value of this Mihrab geometry for 121 sound receiving points on mosque prayer rows is 0.65 dB.

3.7 Umayyad Spain Mihrab geometry SPL analysis

Figure 16 depicts the SPL contours and SPL values for all receiver points obtained by Umayyad Spain Mihrab geometry. As Almoravid, Spanish Umayyad Mihrab has an octagonal shape. The top end of this Mihrab is different from that of Almoravid Mihrab. As shown in **Figure 16**, this slight difference caused an increase of sound energy especially on the left side of the mosque. The highest SPL values are found at the first three prayer rows due to the proximity to the sound source as well as the middle portion of each prayer row (behind Imam position). The yellow-hatched area covers 58% of the mosque floor plan with high SPL values that range from 72 to 60 dB. Moreover, this figure shows the three main sound directions of Mihrab geometry, which are good for rectangular floor plan. **Figure 17** presents SPL average values for prayer rows of Umayyad Spain masjid Mihrab geometry. The average values vary from 71 to 65 dB. The highest SPL average value is at the first prayer row and starts decreasing gradually toward the tenth prayer row. Slight increment in SPL average value is found at the last/eleventh prayer row.

3.8 Mughal Mihrab geometry analysis

Figure 18 illustrates the SPL values of Mughal Mihrab geometry. The SPL values are varying from 49 to 73 dB on the plan reference to SPL scale in **Figure 18**. The

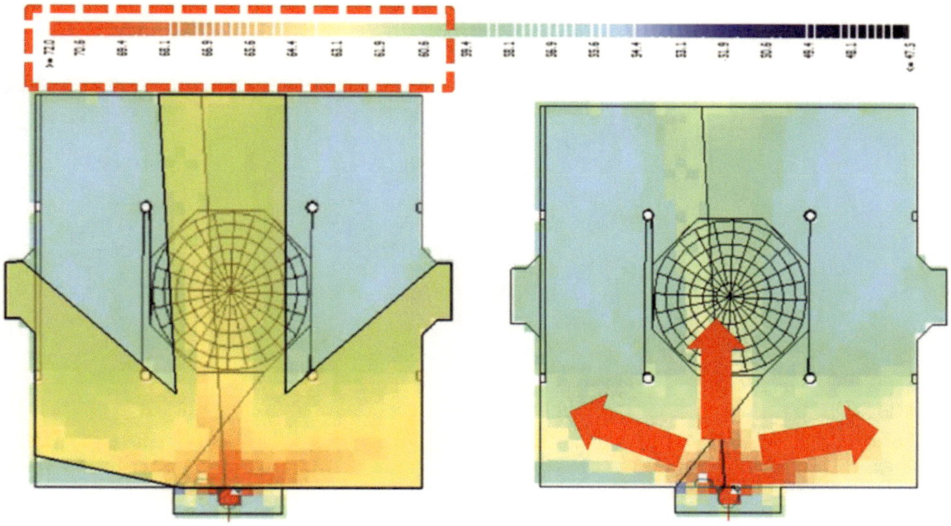

Figure 16.
SPL contour values for Spanish Umayyad Mihrab at 1 kHz.

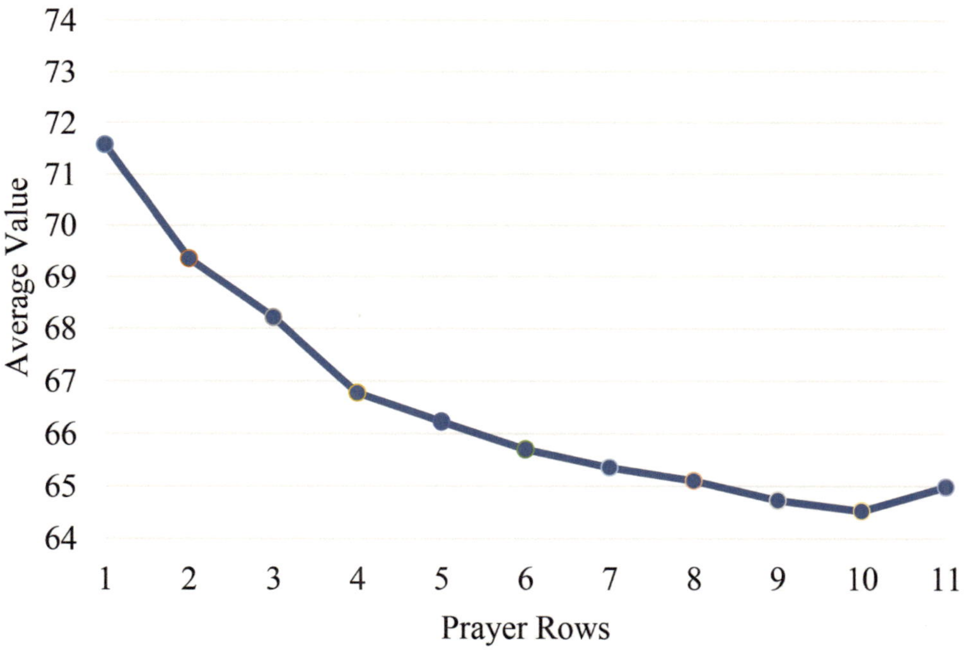

Figure 17.
SPL average values for Umayyad Spain Mihrab prayer rows.

SPL pattern shows nonsymmetrical sound distribution all over the floor plan. The yellow-highlighted area covers 69% of floor plan with high SPL values that range from 73 to 60 dB. Contrary to previous shapes, this shape covers most of floor plan area with high SPL values, which is suitable for most floor plan geometries. **Figure 19** shows SPL average values for prayer rows of Mughal masjid Mihrab geometry. The average values vary from 72 to 66 dB. Besides, the highest SPL average value is received by the first prayer row and then decreases sharply of average values toward the fourth prayer row. There is a slight decrease of average values from the fourth to the seventh prayer row. The seventh, eight, and eleventh

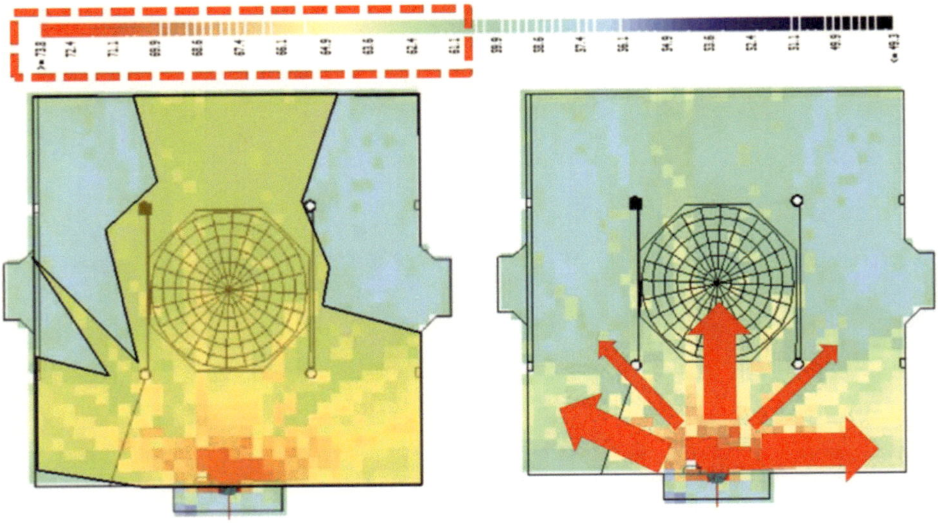

Figure 18.
SPL contour values for Mughal Mihrab at 1 kHz.

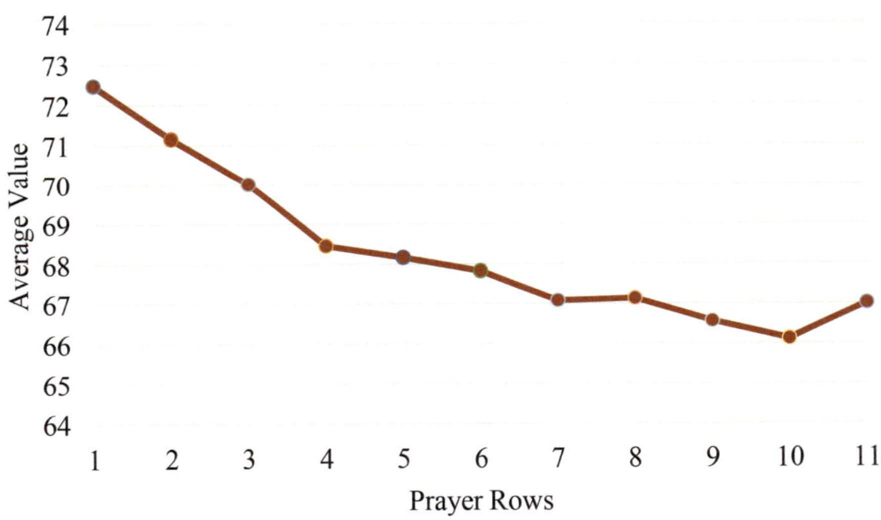

Figure 19.
SPL average values for Mughal Mihrab prayer rows.

prayer rows have the same SPL average values. The lowest average values are found at the tenth prayer row.

3.9 Abbasid Mihrab geometry analysis

Figure 20 shows SPL contour obtained by the Abbasid Mihrab. This Mihrab provides a very high SPL values on all floor area. The SPL values are varying from 49 dB to 73 dB on the plan reference to SPL scale in **Figure 20**. The SPL pattern shows nonsymmetrical sound distribution all over the floor plan. The yellow-highlighted area covers 69% of floor plan with high SPL values that range from 73 to 60 dB. Contrary to previous shapes, this shape covers most of floor plan area with high SPL values, which is suitable for most floor plan geometries.

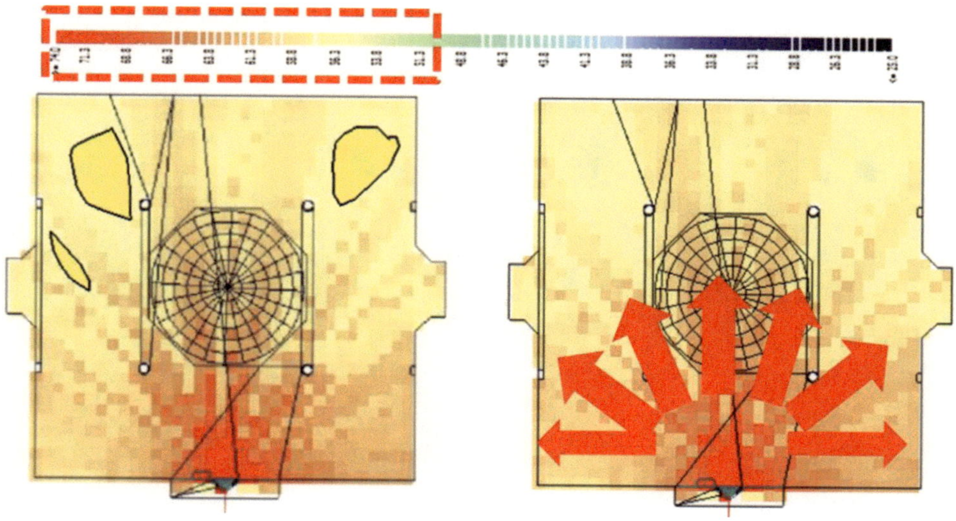

Figure 20.
SPL contour values for Abbasid Mihrab at 1 kHz.

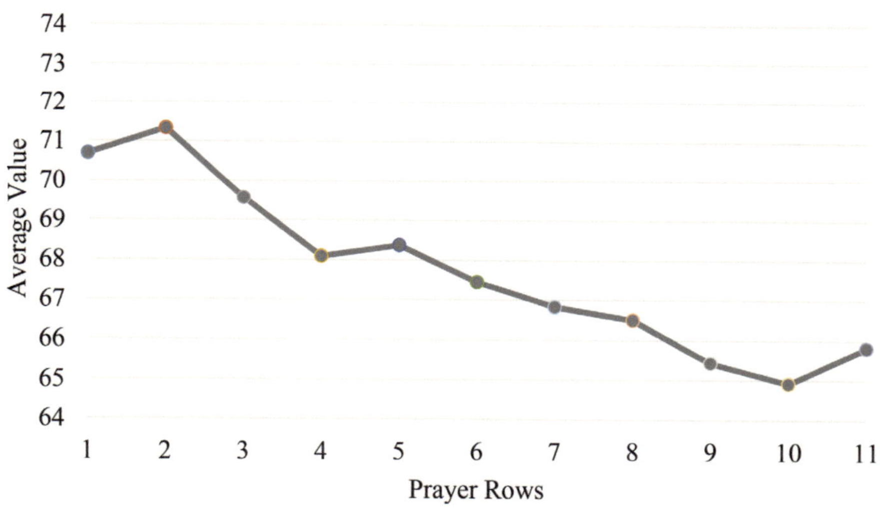

Figure 21.
SPL average values for Abbasid Mihrab prayer rows.

Figure 21 illustrates SPL average values for prayer rows of Abbasid Mihrab geometry. The average values vary from 71 to 65 dB. High and low levels of SPL values are shown from the first to the fifth row. Highest SPL average value is found at the second prayer row where it is higher than the first prayer row. In addition to that, the lowest SPL average value is found in the tenth prayer row. From the sixth to the tenth prayer rows, SPL average values are gradually decreasing to the lowest SPL average value.

3.10 SPL analysis for Tulunid Mihrab geometry

Figure 22 presents the SPL values of Tulunid Mihrab geometry. The SPL values are varying from 49 to 74 dB on the plan reference to SPL scale. It is observed that

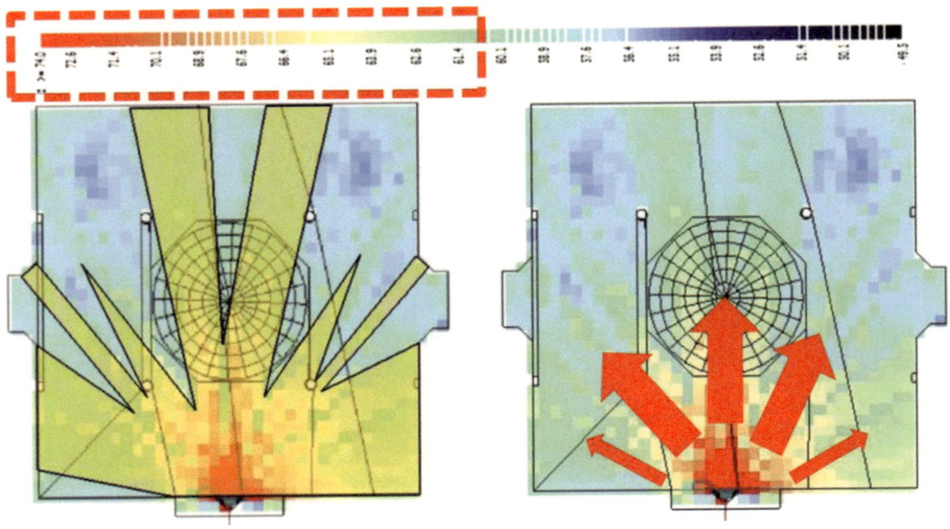

Figure 22.
SPL contour values for Tulunid Mihrab at 1 kHz.

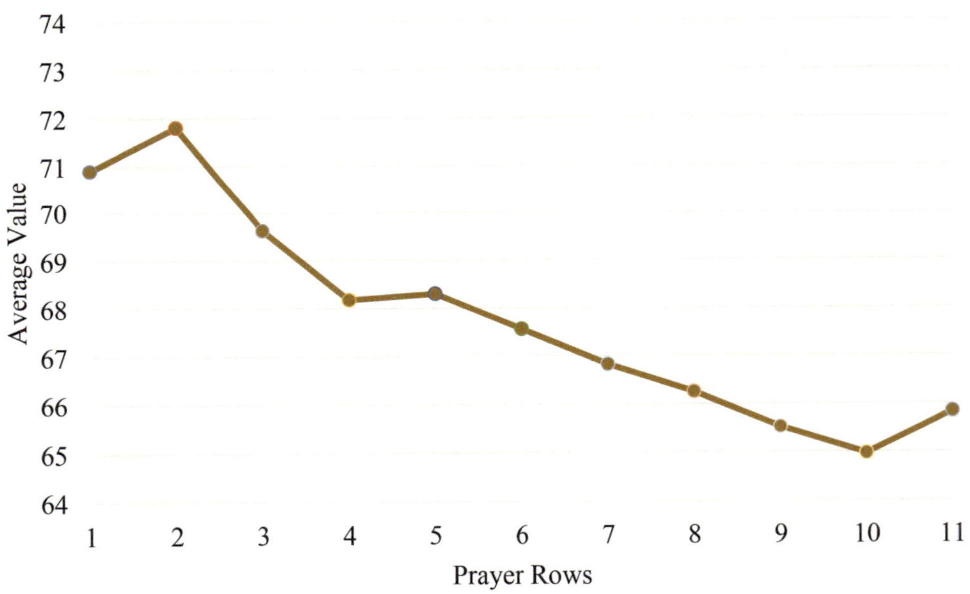

Figure 23.
SPL average values for Tulunid Mihrab prayer rows.

Tulunid Mihrab shape has symmetrical sound distribution and yellow-hatched area covers 57% of Mosque floor plan area with high SPL values that range from 74 to 60 dB. Moreover, Tulunid Mihrab shape is suitable for the square floor plan geometry. **Figure 23** illustrates SPL average values for prayer rows of Tulunid Mosque Mihrab geometry. The average values vary from 72 to 65 dB, and high and low levels of SPL values are shown from the first to the fourth prayer row. Peak of SPL average value is found at the second prayer row where it is higher than the first prayer row. In addition, the lowest SPL average value is found in the tenth prayer row. From the fifth to the tenth prayer row, SPL average values are gradually decreasing to the lowest SPL average value. The average SPL value of this Mihrab geometry for 121 sound receiving points on mosque prayer rows is 0.70 dB.

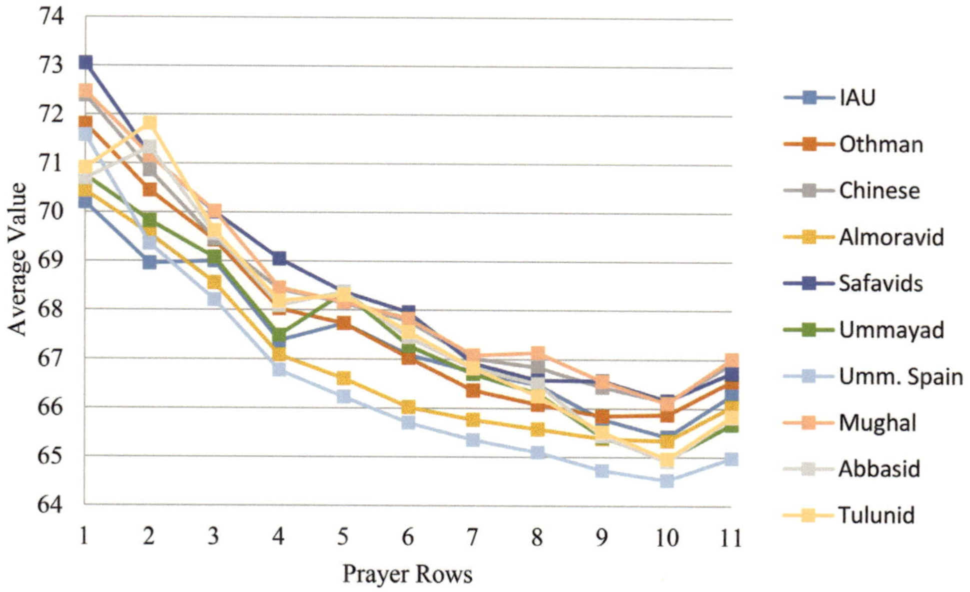

Figure 24.
SPL prayer rows' average values, for all studied Mihrab.

3.11 Overall evaluation

As shown in **Figure 24**, the SPL average values for the Mihrab geometries are studied. In general, the SPL average values show good (greatest) values at the first prayer row, progressively decrease to the tenth prayer row, and then increase to the eleventh prayer row. Moreover, at the first four prayer rows, the Safavid has the highest average values and base case mosque has the lowest. Safavid and Mughal have high SPL average values at the first three rows as well, and the rest of geometries still receive good SPL average values. Furthermore, from the fourth toward the sixth prayer row, the Safavid still receives the highest average value besides Tulunid, Mughal, and Umayyad Mihrab geometries. Almoravid and Umayyad Spain Mihrab geometries have the lowest SPL average values among other geometries in this area. At the same portion, other geometries have higher SPL average values, which are close to the highest average values. From the sixth to the ninth prayer row, the receiving points decrease when we go far from sound source. Mughal Mihrab geometry has the best SPL average values at these rows as well as Chinese, Safavid, and Abbasid geometries. The worst SPL values at these prayer rows are found for Almoravid and Umayyad Spain Mihrab shapes. From the ninth to the eleventh prayer rows, Safavid and Othman receive the highest SPL average values, while the lowest (worst) values are for Tulunid, Abbasid and Umayyad Spain.

In order to clarify further on the acoustic performance of the studied geometries and to rank them accordingly, they are graded based on the total SPL average

SPL Average Values Grades										
Grade	1	2	3	4	5	6	7	8	9	10
SPLValue	85	80	75	70	65	60	55	50	45	40

Table 3.
SPL average value grade key table.

Prayer Rows	Mihrab Geometries									
	Base Case	Othman	Chinese	Almoravid	Safavids	Umayyad	Umayyad Spain	Mughal	Abbasid	Tulunid
1st	70	70	70	70	75	70	70	70	70	70
2nd	70	70	70	70	70	70	70	70	70	70
3rd	70	70	70	70	70	70	70	70	70	70
4th	65	70	70	65	70	70	65	70	70	70
5th	70	70	70	65	70	70	65	70	70	70
6th	65	65	70	65	70	65	65	70	65	70
7th	70	65	65	65	65	65	65	65	65	65
8th	65	65	65	65	65	65	65	65	65	65
9th	65	65	65	65	65	65	65	65	65	65
10th	65	65	65	65	65	65	65	65	65	65
11th	65	65	65	65	65	65	65	65	65	65
Total Grade	50	50	49	52	48	50	52	49	50	49
Rank	3rd	3rd	2nd	4th	1st	3rd	4th	2nd	3rd	2nd

Table 4.
Total SPL average values, grades, and rank of the Mihrab geometries.

values by following the rule presented in **Table 3**. Based on this rule, the SPL average value at each prayer row was assigned the grades which were summed up to obtain the total grade for each geometry. A lower total grade indicates higher SPL average value, which means a good acoustic performance; accordingly, the geometries were ranked by assigning the first rank for the lowest total grade and lower ranks for higher total grades, as tabulated in **Table 4**. It is obvious from **Table 4** that Safavid Mihrab geometry has the highest and best SPL average value, which is the best value for the first prayer row among the study of Mihrab geometries. Chinese, Mughal, and Tulunid Mihrab geometries have the second best performance in the study, and the worst performance is for Almoravid Mihrab geometry with 52 grades.

4. Conclusion

The Mihrab geometry has always been one of the major features in mosque's architecture, which directly affect the sound quality inside the prayer hall. This sound quality has been tested and simulated in a typical mosque within the Abdulrahman Bin Faisal University campus using 10 principal Mihrab shapes as a variable parameter in correlation with the fixed mosque shape and volume. Out of the 10 geometries studied, the Safavid Mihrab geometry can provide the best acoustic performance inside the mosque, followed by Chinese, Mughal, and Tulunid, while the Almoravid geometry shows the worst performance. Thus, the present study has established that the geometry of the Mihrab, as an Islamic architectural feature, can have a direct positive or negative impact on the sound quality within a prayer hall.

Author details

Hany Hossameldien[1*] and Abdulrahman Abdullah Alshawan[2]

1 Department of Building Engineering, College of Architecture and Planning, Imam Abdulrahman Bin Faisal University, Dammam, Saudi Arabia

2 Saudi Basic Industries Corporation (SABIC), Dammam, Saudi Arabia

*Address all correspondence to: hehassam@iau.edu.sa

References

[1] Topaktaş İL. Acoustical properties of classical ottoman mosques–simulation and measurements (Doctoral Dissertation). The Middle East Technical University; 2003. Retrieved from: https://etd.lib.metu.edu.tr/upload/920116/index.pdf

[2] Zühre SG, Mehmet Ç. Impact of design decisions on acoustical comfort parameters: Case study of Doğramacızade Ali Paşa Mosque. Applied Acoustics. 2013;**74**:834-844

[3] Carvalho APO, Freitas CPT. Acoustical characterization of the central mosque of Lisbon. Forum Acusticum. 2011;**27**:43-55, Aalborg

[4] El-khateeb AA, Ismail MR. Sounds from the past: The acoustics of sultan Hassan mosque and madrasa. Building Acoustics. 2007;**14**(2):109-132

[5] Zühre S, Yilmazer S. The acoustical characteristics of the Kocatepe mosque in Ankara, Turkey Zühre. Earthscan. 2006;**51**(1):21-30. DOI: 10.3763/asre.2008.5104

[6] Carvalho POA, Monteiro CG. Comparison of the acoustics of mosques and catholic churches. 16th International Congress on Sound and Vibration, Kraków, Poland, 5–9 July 2009. Laboratory of Acoustics, Faculty of Engineering, Univ. of Porto, 4200-465 Porto, Portugal

[7] Fausti P, Pompoli R, Prodi N. Comparing the acoustics of mosques and Byzantine churches. EU Contract ICA3-CT-1999-00007. Conservation of the Acoustical Heritage and Revival of Sinan's Mosques Acoustics-CAHRISMA, 2000–2003. Publisher: Dipatimento di Ingegneria Engineering, Università di Ferrara Via Saragat 1, 44100 Ferrara, Italia

[8] Carvalho AP, Amado JD. The Acoustics of the Mekor Haim Synagogue, Portugal. internoise 2011, Osaka, Japan, September 4–7, 2011. PORTUGAL, Laboratory of Acoustics, Faculty of Engineering, University of Porto 4200-465 Porto

[9] de Sant'Ana DQ, Zannin PHT. Acoustic evaluation of a contemporary church based on in situ measurements of reverberation time, definition, and computer-predicted speech transmission index. Building and Environment. 2011;**46**:511-517

[10] Zerhan K, Sevda E. Comparison of the acoustical properties of an ancient and a recent mosque. In: CAHRISMA Research Project. Faculty of Architecture, Istanbul, Turkey: Yýldýz Technical University; 2002. pp. 43-55

[11] Abdou AA. Comparison of The Acoustical Performance of Mosque Geometry Using Computer Model Studies. Eighth International IBPSA Conference, Eindhoven, Netherlands, August 11–14, 2003. Saudi Arabia: King Fahd University of Petroleum and Minerals

[12] Ahmad Y, Din NC, Othman R. Mihrab design and its basic acoustical characteristics of traditional vernacular mosques in Malaysia. Journal of Building Performance. 2013;**4**(1). Department of Architecture, University of Malaya, Malaysia. ISSN: 2180-2106

[13] Utami SS. An acoustical analysis of domes coupled to rooms, with special application to the Darussholah Mosque (Master Thesis). East Java, Indonesia: Brigham Young University; 2005 Retrieved from: http://scholarsarchive.byu.edu/cgi/viewcontent.cgi?article=1654&context=etd

[14] Kayili M. Acoustic Solutions In Classic Ottoman Architecture. Foundation for Science Technology and Civilisation. Gazi University, Faculty of

Architecture & Engineering, Ankara, Turkey. 2005. Publication ID: 4087. Retrieved from: http://www.fstc.co.uk

[15] Cunha IB, Smiderle R, Bertoli SR. Influence of sound reinforcement system on acoustic performance in a Catholic Church. International Symposium on Room Acoustics, June 9–11, 2013, Toronto, Canada. Campinas, Brazil, Univerity of Campinas

[16] Orfali W. room acoustic and modern electro acoustic sound system design during constructing and reconstructing mosques (Doctoral Dissertation). Berlin: Technischen Universität; 2007. Retrieved from: http://citeseerx.ist.psu.edu/viewdoc/download?doi=10.1.1.601.4879&rep=rep1&type=pdf

[17] Hossam Eldien H, Al Qahtani H. The acoustical performance of mosques main prayer hall geometry in the eastern province, Saudi Arabia. Société Française d'Acoustique. Acoustics 2012, Ap. 2012, Nantes, France. <hal-00810652>

[18] Abdou AA. Measurement of acoustical characteristics of mosques in Saudi Arabia (Research). 2003 Acoustical Society of America. King Fahd University of Petroleum and Minerals, Dhahran 31261, Saudi Arabia; 2002. DOI: 10.1121/1.1531982, PACS numbers: 43.55.Br, 43.55.Gx, 43.55.Jz, 43.55.Mc. Retrived from: https://www.ncbi.nlm.nih.gov/pubmed/12656385

[19] Catholic Diocese of Columbus, Ohio, USA. Acoustics for Churches and Chapels, A Summary of Concerns, Guiding Principles, and Recommendations for Acoustics for the Construction and Renovation of Worship Spaces. 2003. Retrieved from: http://www.colsdioc.org/Portals/0/Departments/LIT/Documents/GuideAcoustics.PDF

[20] Francesco M, Ettore C, Antonio C, Paola R. Guidelines for acoustical measurements in churches. Applied Acoustics. 2009;**70**:378-388

[21] Ismail MR. A parametric investigation of the acoustical performance of contemporary mosques. SciVerse ScienceDirect. 2013;**2**:30-41

[22] Rosniza O. Mihrab Design and Ornamentation of Selected Mosques in Malaysia (Doctoral Dissertation). Universiti Malaya; 2011 Retrieved from http://studentsrepo.um.edu.my/3345/